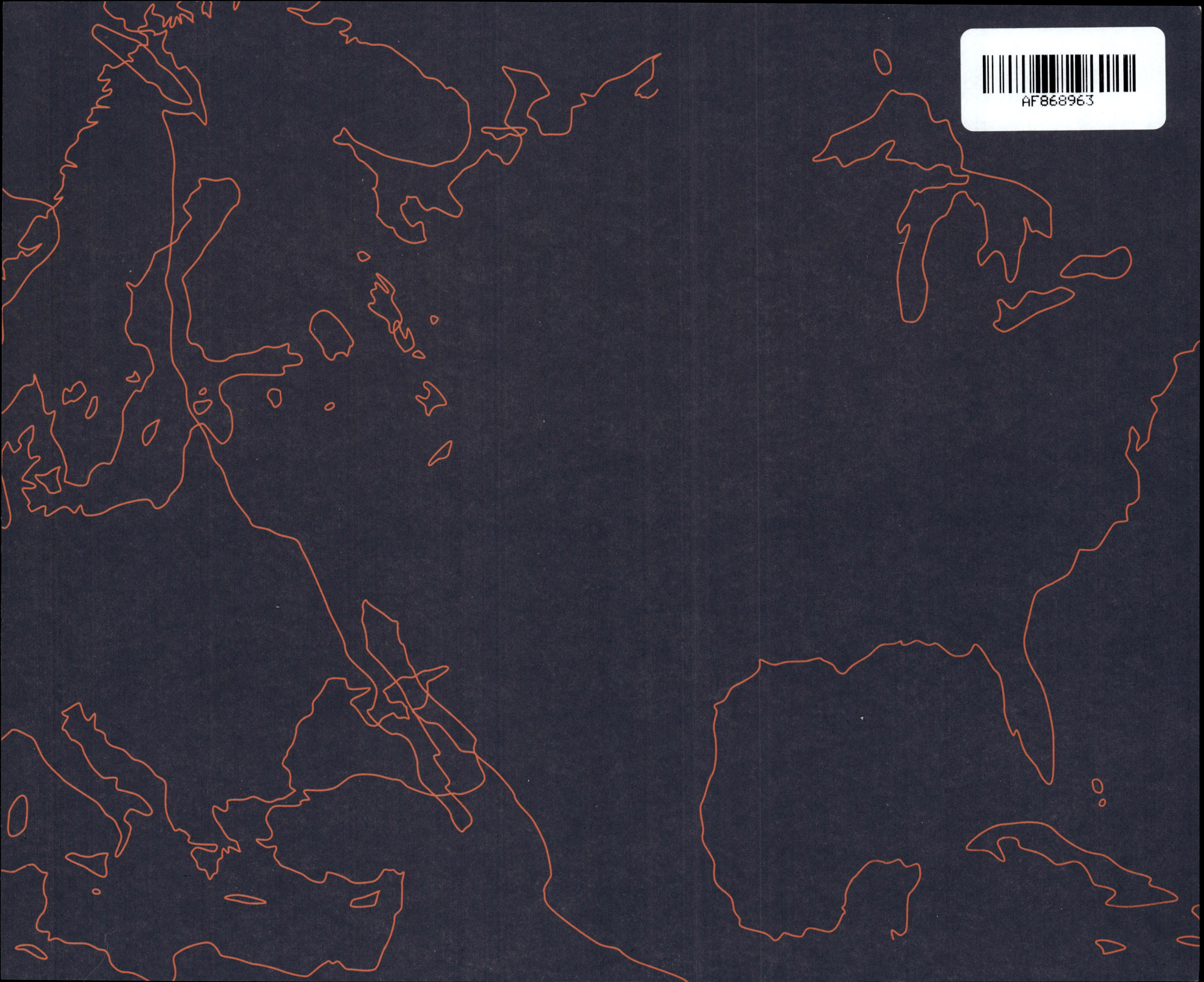

IN EXTREMIS

IN EXTREMIS

Landscape Into Architecture

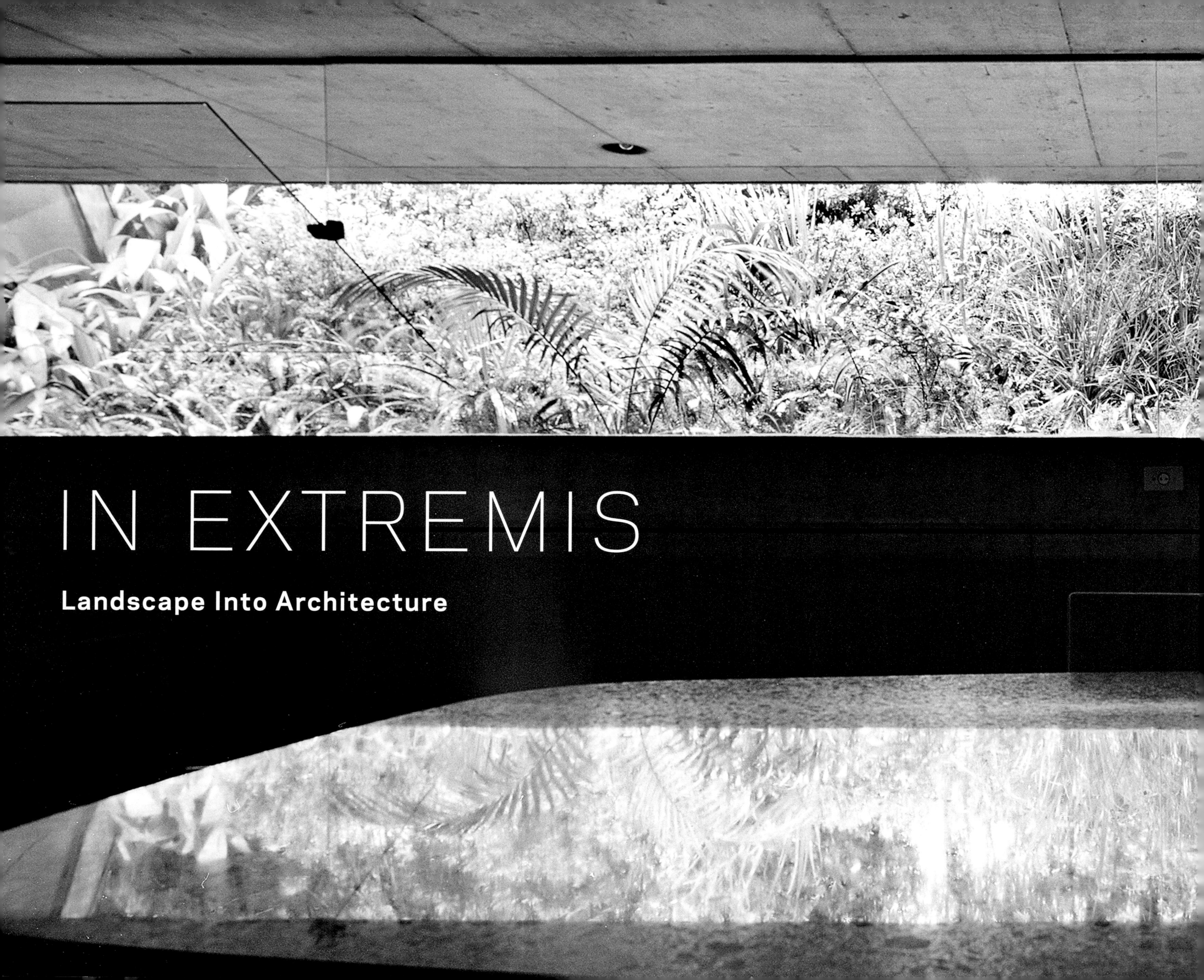

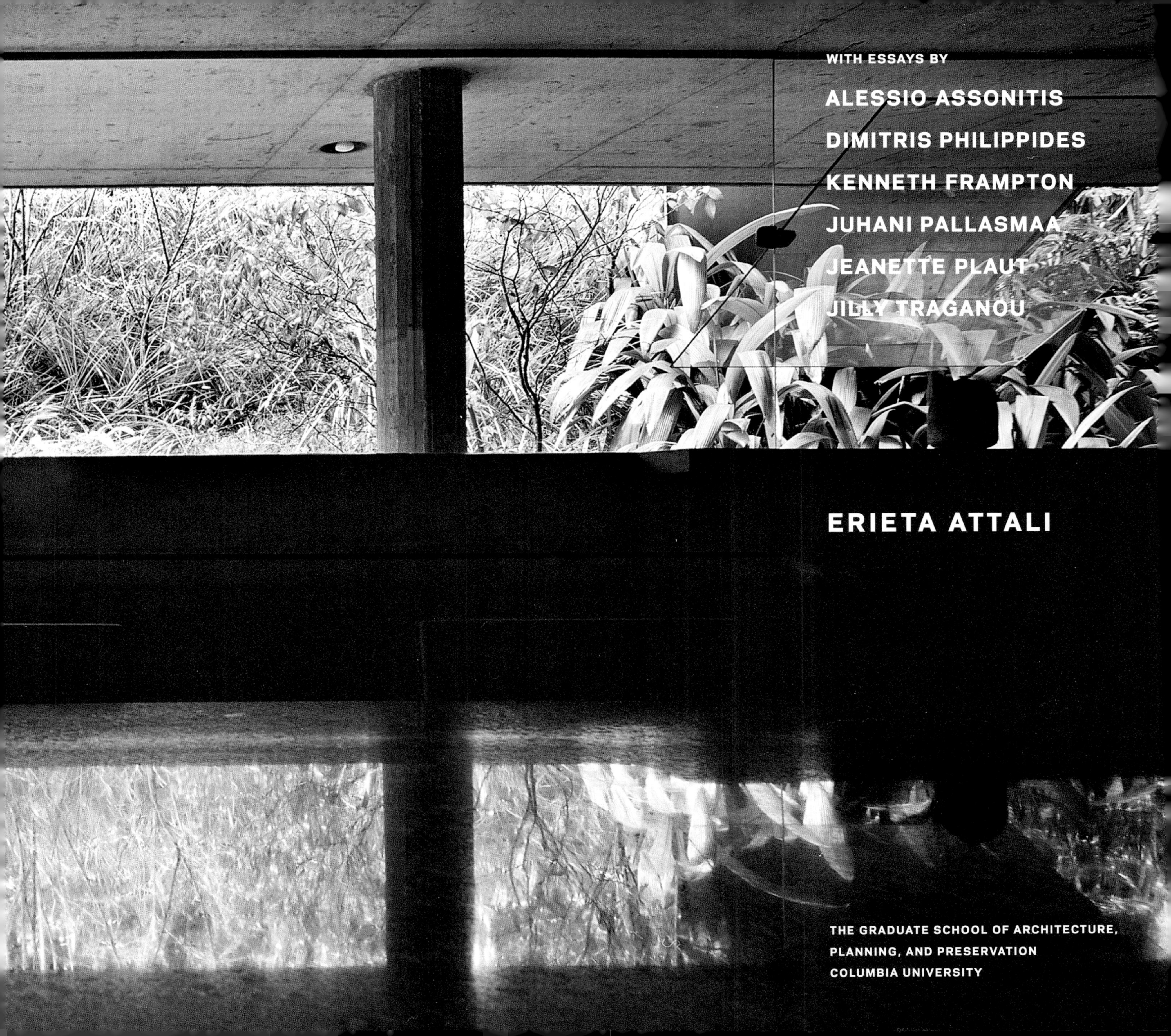

WITH ESSAYS BY

ALESSIO ASSONITIS

DIMITRIS PHILIPPIDES

KENNETH FRAMPTON

JUHANI PALLASMAA

JEANETTE PLAUT

JILLY TRAGANOU

ERIETA ATTALI

THE GRADUATE SCHOOL OF ARCHITECTURE,
PLANNING, AND PRESERVATION
COLUMBIA UNIVERSITY

Published by
The Graduate School of Architecture, Planning, and Preservation
Columbia University
1172 Amsterdam Ave.
409 Avery Hall
New York, NY 10027

Visit our website at www.arch.columbia.edu/publications

Printing and Binding by Everbest Printing, China
First Edition

COVER PICTURE
dNR Architects / Max Nuñez - Nicolas del Rio
Chalet 7, Portillo, Chile

The book was produced with the generous support of Diathlasis Lighting, Athens.

diathlasis architectural lighting

This book has been produced through the Office of the Dean, Mark Wigley,
with production coordination by Craig Buckley, Office of Print Publications.

GRAPHIC DESIGN
HvADesign, NY (Henk van Assen with Anna Zhang)
COPY EDITING
Craig Buckley and Pollyanna Rhee

Library of Congress Cataloging-in-Publication Data
Attali, Erieta, 1966–
In extremis : landscape into architecture / Erieta Attali ; with essays by
Alessio Assonitis ... [et al.]. — 1st ed.
p. cm.

Includes index.

ISBN 978-1-883584-67-2 (alk. paper)
1. Architecture, Modern—21st century. 2. Building sites. I. Assonitis, Alessio. II. Title.
III. Title: Landscape into architecture.

NA687.A88 2010
724'.7—DC22

2010040077

Table of Contents

For Clio and all my friends around the world

I am enormously grateful to the following people and institutions for their tremendous support:

Ferdy Carabott, colleague and dear friend, for his invaluable technical expertise and generous sharing of knowledge and experience in photography over the years; Henk van Assen for the long hours of work he heartily contributed for the design of this book; Craig Buckley, director of publications at GSAPP, Columbia University, for overseeing the project and keeping a keen eye on details; Juhani Pallasmaa, Alessio Assonitis, Jilly Traganou, Jeannette Plaut, Dimitris Philippidis and Kenneth Frampton for their insightful essays; Karin Skousbøll for showing unequaled benevolence and patience while working on our common project in the Nordic Countries; ETH Studio Monte Rosa / Bearth & Deplazes Architects, Zurich, Switzerland, for making my journey to Monte Rosa Hut possible; David Hinkle, associate dean for administration at GSAPP, Columbia University, for his support over the last ten years; Agapitos Legakis, John Petrochilos, Stefan Bischof, and all my other assistants around the world for their unstinting help during shootings in many parts of the world; Richard Posamentir and his family, Maruja Caceres, Sotiria Tzitzi, Adriani Rogakos, Ephtihia & Argyris Papadopoulos, (all) for being real friends in good times and bad; and of course, to the owners of the houses featured in this collection, for letting a "stranger" temporarily upset their daily lives, thus showing the true meaning of hospitality.

Kenneth Frampton, Mark Wigley, Bernard Tschumi, Kengo Kuma, Angelo Bucci, Juhani Pallasmaa, and all the architects whose work is featured in this book have personally, and through their work, been valued stimuli and a tremendous inspiration. I am forever indebted to them.

In addition, special thanks must go to the following institutions:

The Graduate School of Architecture, Planning, and Preservation, Columbia University, New York
The Danish Arts Council, Copenhagen
The Fulbright Foundation, Athens
The Graham Foundation for Advanced Studies in the Fine Arts, Chicago
The Japan Foundation, Tokyo
The Dreyers Foundation, Copenhagen
The Royal Danish Academy of Arts, Copenhagen
The Faculty of Architecture of the Swiss Federal Institute of Technology, Zurich

Finally, I would like to express my gratitude to Panayiotis Ioannidis and his highly skilled staff at Diathlasis Architectural Lighting without whose generous support this book could not have been made possible.

I am thankful to all of you who have offered your support so generously over the years and have helped me to make this work possible.

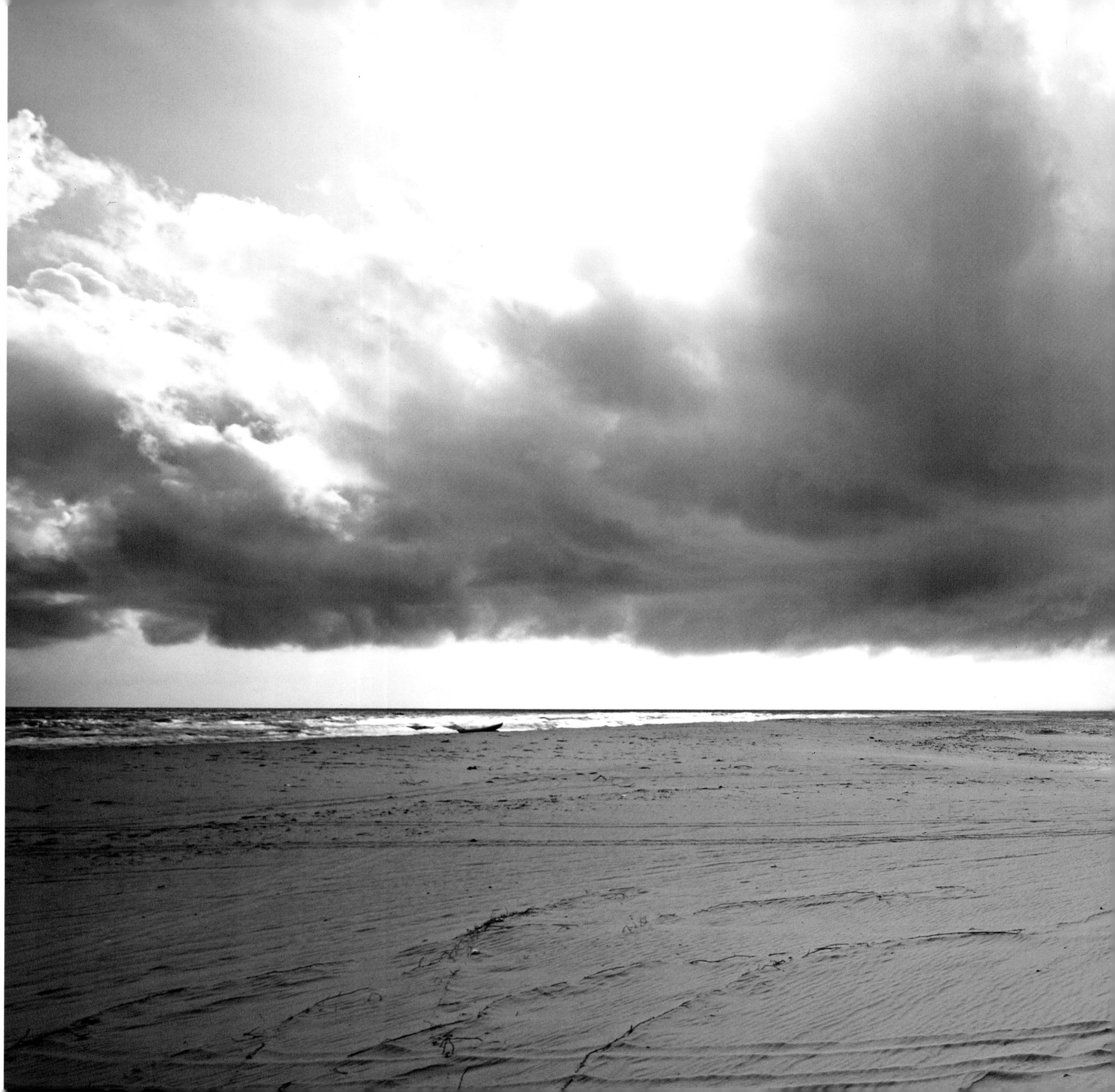

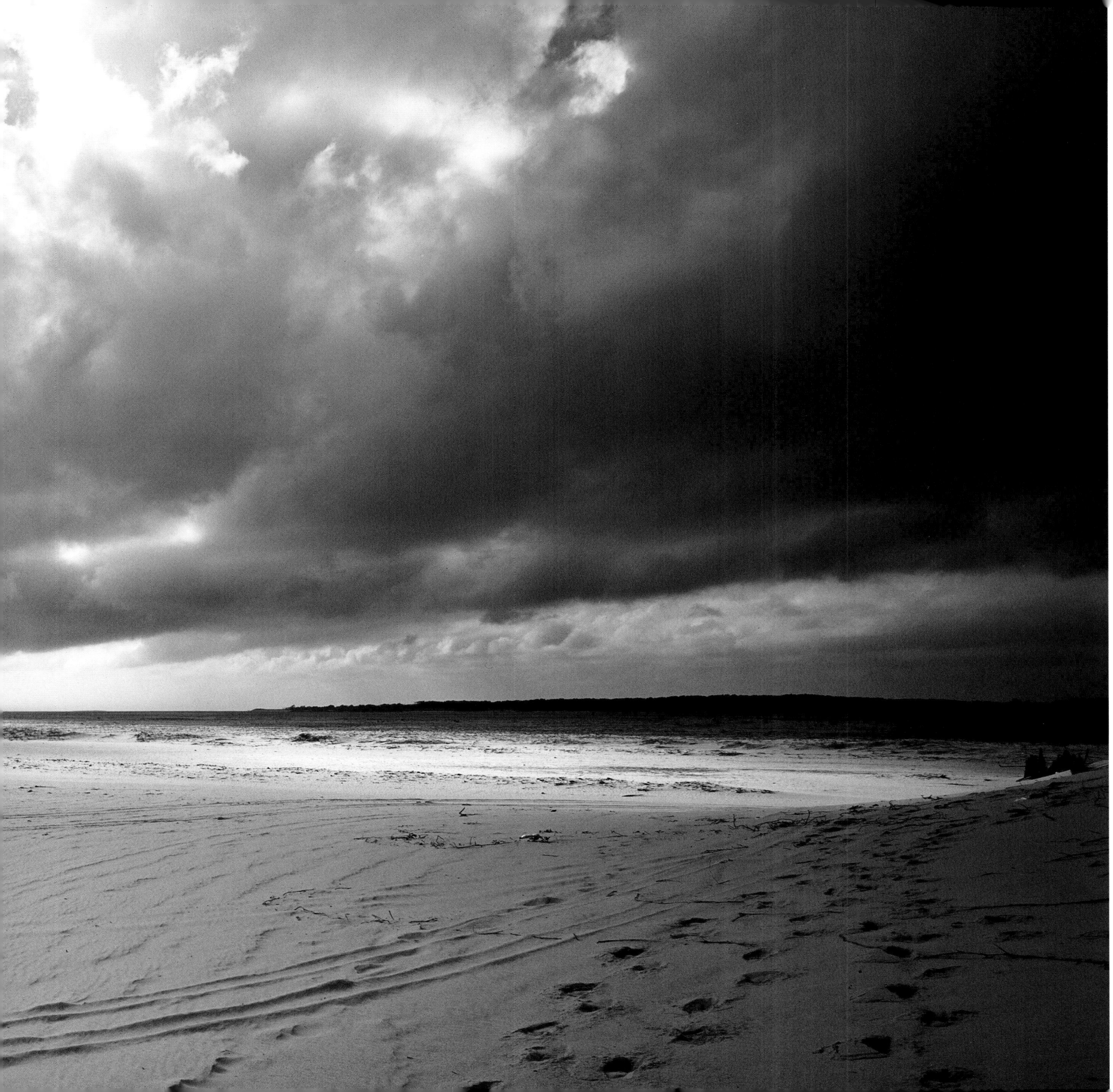

JUHANI PALLASMAA

Landscapes of the Mind

The Inner Space of The World

WE TEND TO THINK OF LANDSCAPES as external settings for our life and architectural structures. However, there are no external settings for us to settle in, as our perceptions and awareness unavoidably imply an exchange; when we enter a space, the space enters us. Physical and geographical landscapes are also mindscapes. We are not living on a given *mise-en-scene,* as the world is of our own making, to some degree materially, but in its entirety in terms of its mental implications. "I am the space, where I am."[1] Noël Arnaud, the poet, confesses, whereas Maurice Merleau-Ponty, the philosopher, reasons: "The world is wholly inside, and I am wholly outside of myself."[2] He calls this process "intertwining" or "chiasma" implying a simultaneous co-existence of the world and the Self within the other without causal or temporal precedence.[3] Rainer Maria Rilke, the master poet, uses the beautiful notion of *Weltinnenraum,* the interior space of the world in reference to the experienced, internalized and embodied world—the landscape of one's intimate domicile and mind.[4]

Erieta Attali's horizontally stretched photographs of buildings in their settings immediately invoke this fused unity, or singularity. The format emphasizes the horizon line, although it may not be visible, and makes the landscape appear endlessly continuous and without boundaries. Attali's landscapes are sublime, bare and powerful, and they address the skeleton and the skin as much as the eyes.

Her photographs record some of the most extreme locations of human dwelling. In these baren landscapes of isolation and solitude, merciless heat and cold, wind and rain, the architectural structures project a sense of arrival, destination and comfort. In Martin Heidegger's words, they express our "coming into the nearness of distance."[5]

The house tames distance, heat and cold, it dims the light, and silences the rage of the storm. The house provides us with our second skin and re-calibrated senses; the house is an instrument that readjusts our vision, hearing and haptic sense. We cannot think clearly in the wild and open nature. We need the focusing device of a constructed room with its ordering geometry and intimate resonance with the body in order to give directionality and determination to our thoughts. An open landscape makes our thoughts stray and hover. As Gaston Bachelard asserts, "[T]he chief benefit of the house [is that] the house shelters daydreaming, the house protects the dreamer, the house allows one to dream in peace."[6]

Buildings are set in the landscape but the landscape is equally mirrored in architecture. Profound architecture always enters a respectful, humble, but at the same time, courageous and proud dialogue with the setting. Paradoxically, the house expresses permanence (albeit, temporary and futile) while the landscape projects the changes of the seasons and hours of the day. The house constitutes the static gnomon on the face of the dynamic sundial of the landscape. As Adrian Stokes remarks, "the hesitancy of water reveals architectural mobility."[7] The house gives the landscape its point of gravity and focus, while the landscape clarifies and amplifies the geometric voice of the house. This is a timeless dialogue and double caress; the landscape embraces the house while the house caresses the landscape.

In our frustrated and misguided time, architecture is usually regarded as an aestheticized object. Yet, profound architecture is never an object, as it is always relational and a mediation. Architecture is a verb which frames what exists and provides a horizon for understanding. For understanding what? Understanding and internalisation of the human condition, the enigma of existence and of our shared history and destiny with the world. Every profound building mediates between the world and ourselves and settles and inhabits us in the "flesh of the world," to use the wonderful notion of Merleau-Ponty. This is a form of existential understanding that does not call for concepts, words or theories. As Jean-Paul Sartre points out; "Understanding is not a quality coming to human reality from outside, it is its characteristic way of existing."[8] True architecture does not require any explanation, not to speak of apology. It ties the strands of here and beyond, matter and image, usefulness and futility, perception and imagination, into a unity that is lived rather than understood. The true meaning of architecture is an existential and embodied wonder that directly articulates our sense of being and Self. We do not live separately in physical and material, mental and spiritual, worlds. These faculties and dimensions are fully fused in the human existential experience.

We tend to think that architecture is solely a vehicle and instrument to confront space. Yet, architectural structures are equally importantly instruments for the domestication of time. We cannot exist mentally in a measureless time. Physical space and natural time have to be scaled to human measures in order to be grasped by our perceptions and understanding. The most adorable products of architecture create a specific center of gravity and meaning, a focal point from which "the world appears complete and right," to use the touching description by which Pierre Teilhard de Chardin characterizes the magical locus of perfection which he calls "Omega."[9]

There is an unconscious correspondence between the landscape and the human body, in the same way that the house and the body are reversible metaphors. We exist in the landscape as embodied beings sharing with it its very flesh. "Our human landscape is our autobiography, reflecting our tastes, our values, our aspirations and even our fears, in tangible, visible form. We rarely think of landscape that way, and so the cultural record we have written in the landscape is liable to be more truthful than most autobiographies because we are less self-conscious about how we describe ourselves," Pierce F. Lewis writes.[10]

In addition to being a testimony of the deepest qualities of culture, or of their absence, ordinary landscapes expose and externalise our inner mental landscapes, the landscapes of our soul. "In the fusion of place and soul, the soul is as much a container of place as place is a container of soul, and both are susceptible to the same forces of destruction," Robert Pogue Harrison, the literary scholar, asserts. Indeed, every act of construction contains an element of destruction; terrain and vegetation are violated, daylight and sight conditions altered, and the virginity of the natural landscape is lost forever. As Paul Valéry appropriately points out, "Destroying and constructing are equal in importance, and we must have souls for the one and the other [...]."[11]

Irresponsible construction creates irreparable wounds and scars on the face of Mother Earth, whereas responsible and sensitive architecture caresses her features, and underlines her dynamics and characteristics. Sensitive architecture even has the power to mend and cure landscapes, both natural and man-made, that have been violated by thoughtless and senseless acts of our fellowmen. Encountering a scene of brutal force, carelessness and lack of aesthetic sensibilities in the man-made landscape is a moment of looking deep into the troubled soul of man. When a culture loses its sense and desire for beauty, it has already lost its joy of life and sense of optimism. This culture has lost its ethical judgement and is already on its way towards self-destruction.

Architecture creates constructed and lived metaphors of the world and human existence. Like all profound works of art, meaningful architectural works are microcosms, complete and autonomous constructed and abstracted worlds of their own. Great buildings are universes, worlds within the world. Yet, they are in respectful and invigorating dialogue with their settings; together with their setting they create a symphonic relationship. "[H]ave you not noticed, in walking about this city, that among the buildings with which it is peopled, certain are *mute;* others *speak;* and others, finally—and they are the most rare—*sing*?," Paul Valéry asks.[12]

A true piece of architecture always enhances, celebrates, clarifies and strengthens the reading of the landscape and gives it specific meanings. The two are intertwined; the landscape frames architecture from outside, while the architectural structure frames the landscape from within. Erieta Attali's photographs convey convincingly and poetically this essential intertwining and internal dialogue. In her images the landscape empowers the architectural structure whereas the building underlines the sublime beauty of the setting. This relationship of the landscape and architecture is an erotic one. As in an amorous relationship, the autonomy and separateness of the other is respected while protecting and supporting the other's vulnerability.

—In the prairies of Illinois, September 2010

1 As quoted in Gaston Bachelard, *Poetics of Space* (1958), Beacon Press, Boston, 1969, p. 137.

2 Maurice Merleau-Ponty, *The Phenomenology of Perception,* trans. Colin Smith, Routledge and Kegan Paul, London, 1962, p. 407.

3 Maurice Merleau-Ponty, "The Intertwining – The Chiasm", *The Visible and the Invisible,* ed. Claude Lefort, Northwestern University Press, Evanston, Illinois, 1992.

4 As referred to in Liisa Enwald, editor, "Lukijalle" [To the Reader], Rainer Maria Rilke, *Hiljainen taiteen sisin: kirjeitä vuosilta 1900-1926* [The silent innermost core of art: letters 1900-1926], TAI-teos, Helsinki, 1997, p. 8.

5 As quoted in John Berger, *Hold Everything Dear; Dispatches or survival and resistance,* Vintage Books, New York, 2007, p. 144.

6 Bachelard, op. cit., p. 6.

7 Adrian Stokes, "Prologue: at Venice," *The Critical Writings of Adrian Stokes,* vol. II, Thames and Hudson, Plymouth, 1978, p. 88.

8 Jean-Paul Sartre, *The Emotions: An Outline of a Theory* (1939), Carol Publishing Co., New York, 1993, p. 9.

9 Pierre Teilhard de Chardin, *The Phenomenon of Man,* Harper Collins Publishers, New York, 2008.

10 Peirce F. Lewis, "Axioms for Reading the Landscape", in D. W. Meining, editor, *The Interpretation of Ordinary Landscapes: Geographical Essays,* Sunstone Press, Santa Fe, 1975, p. 8.

11 Paul Valéry, "Eupalinos, or The Architect", Paul Valéry: *Dialogues,* Pantheon Books, New York, 1956, p. 70.

12 Paul Valéry, op. cit., p. 83

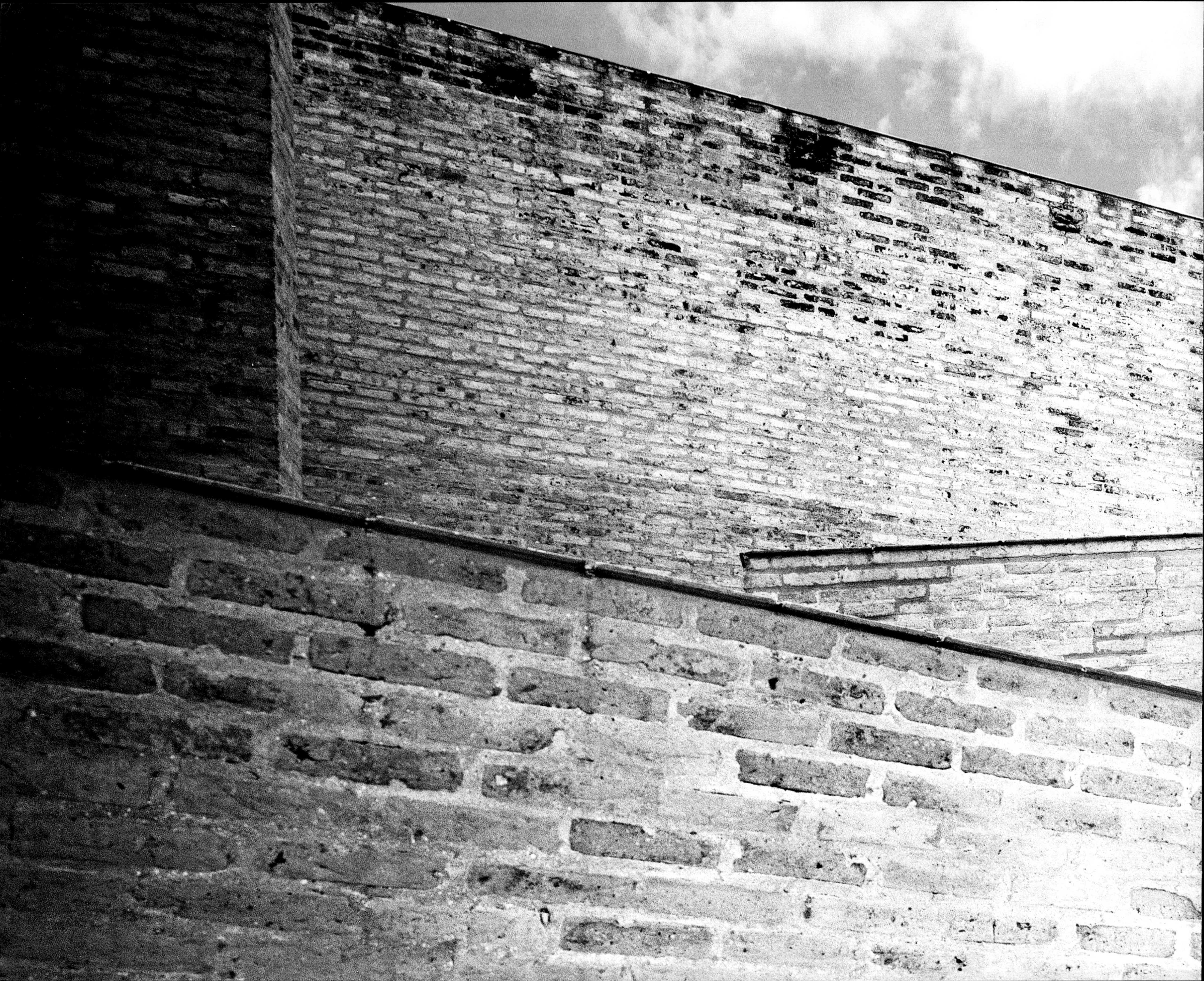

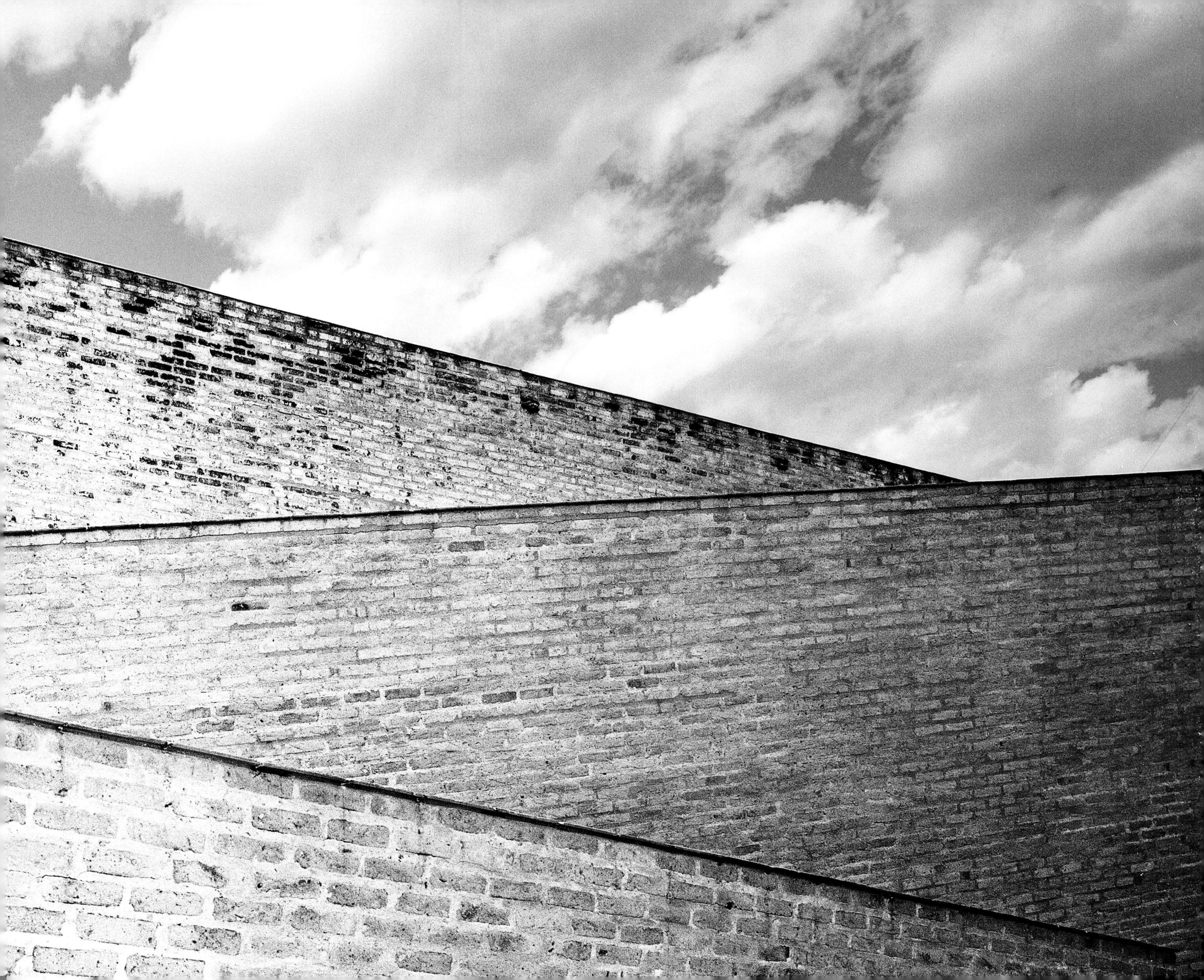

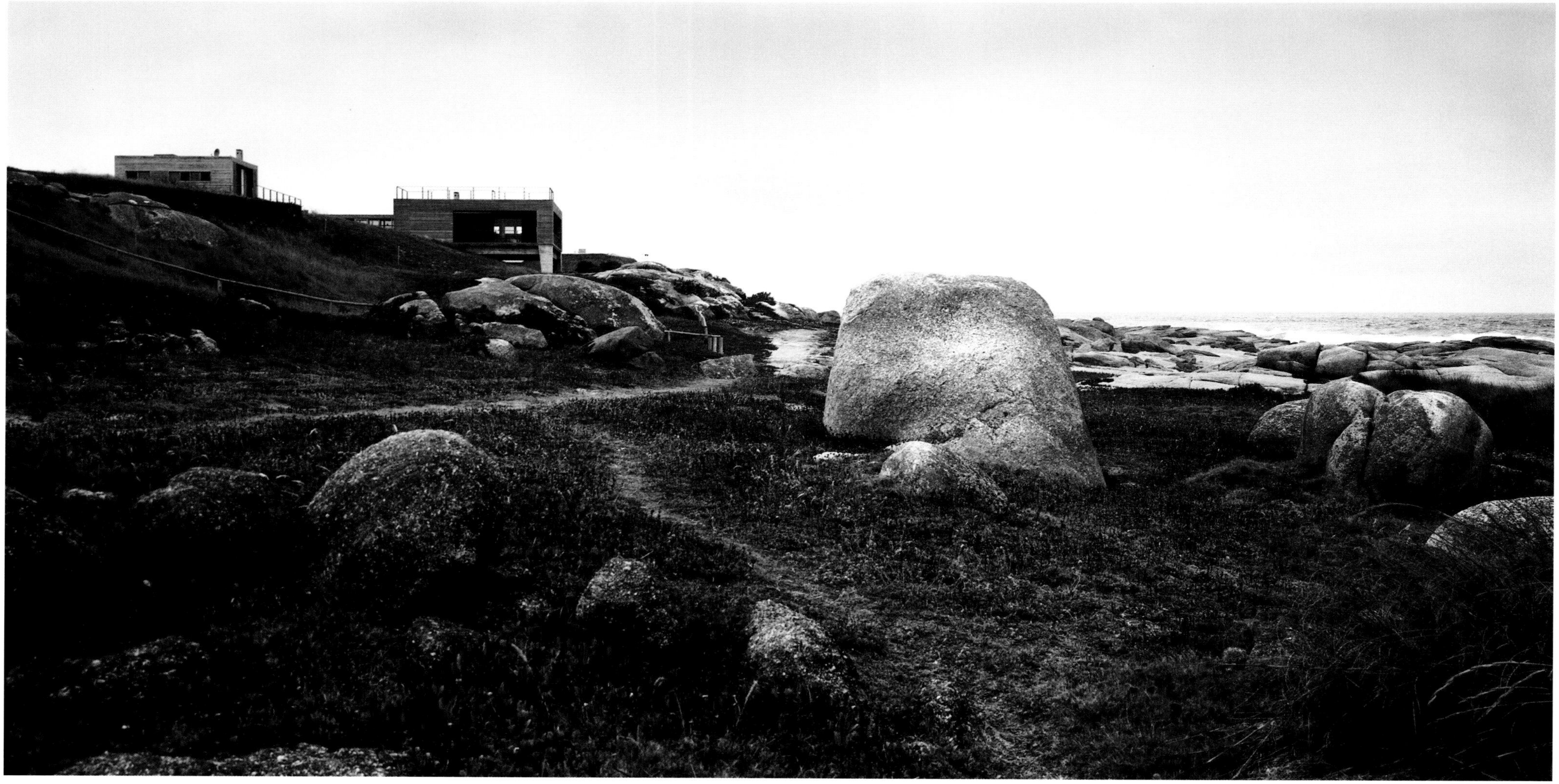

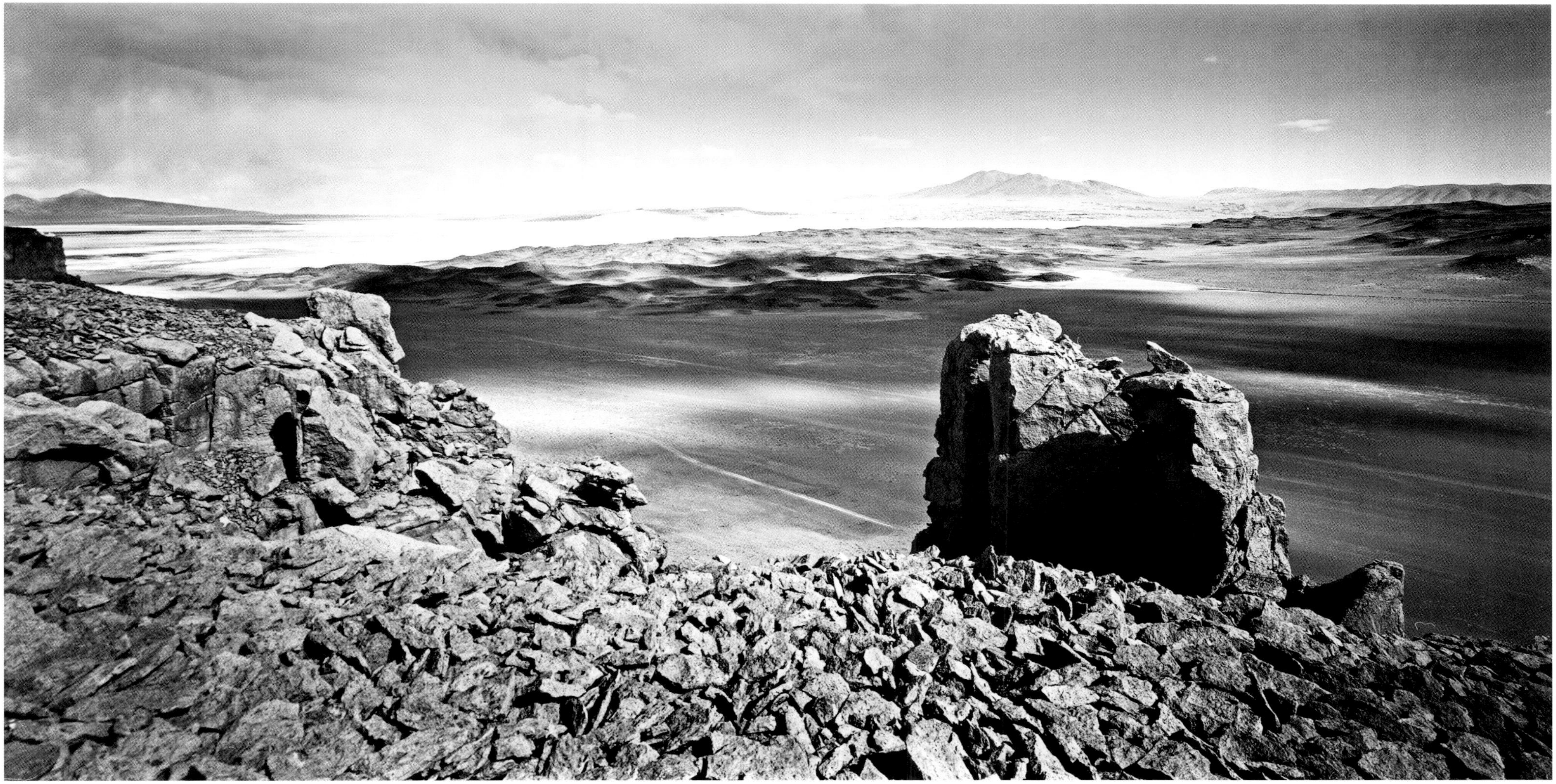

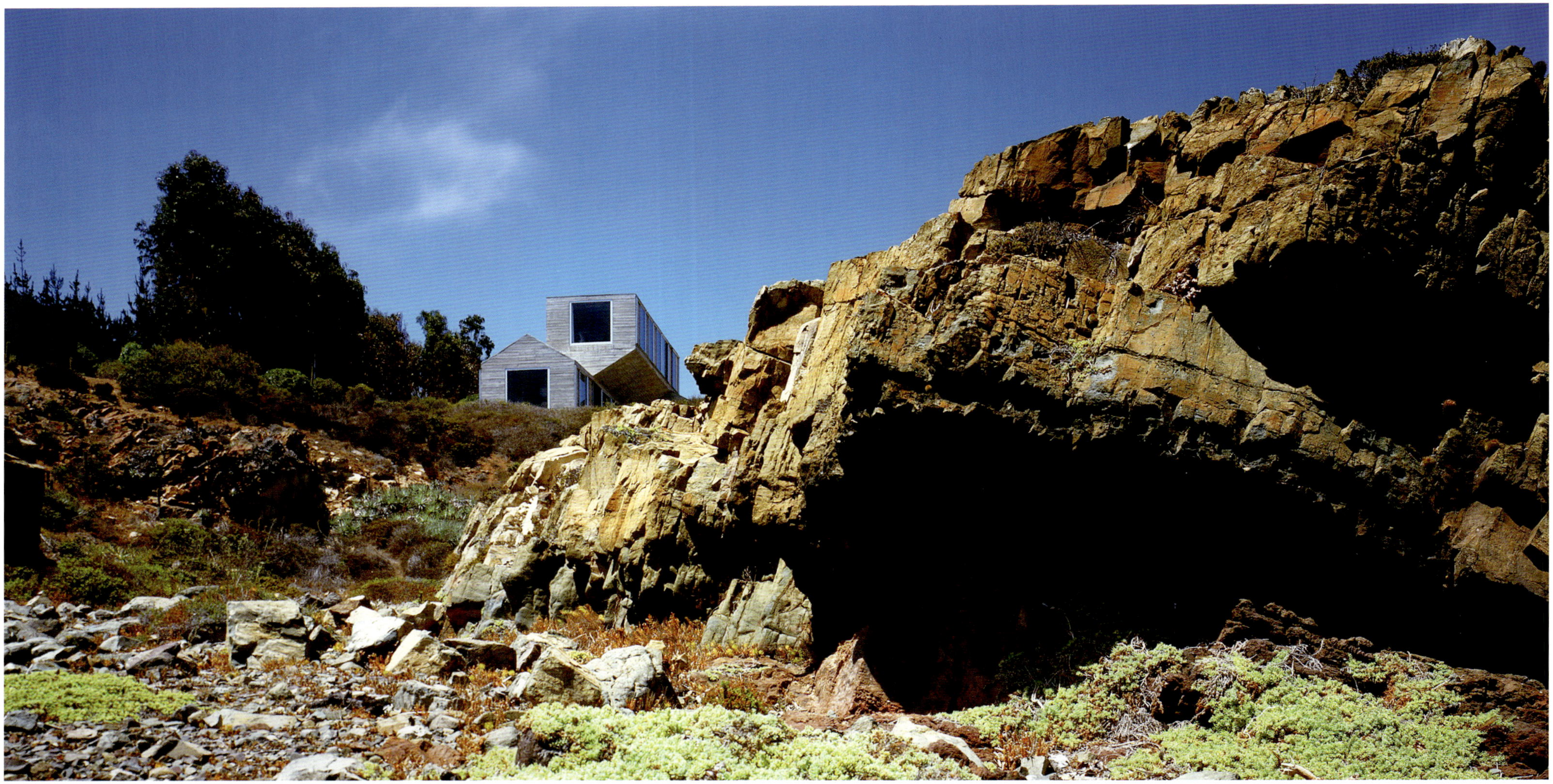

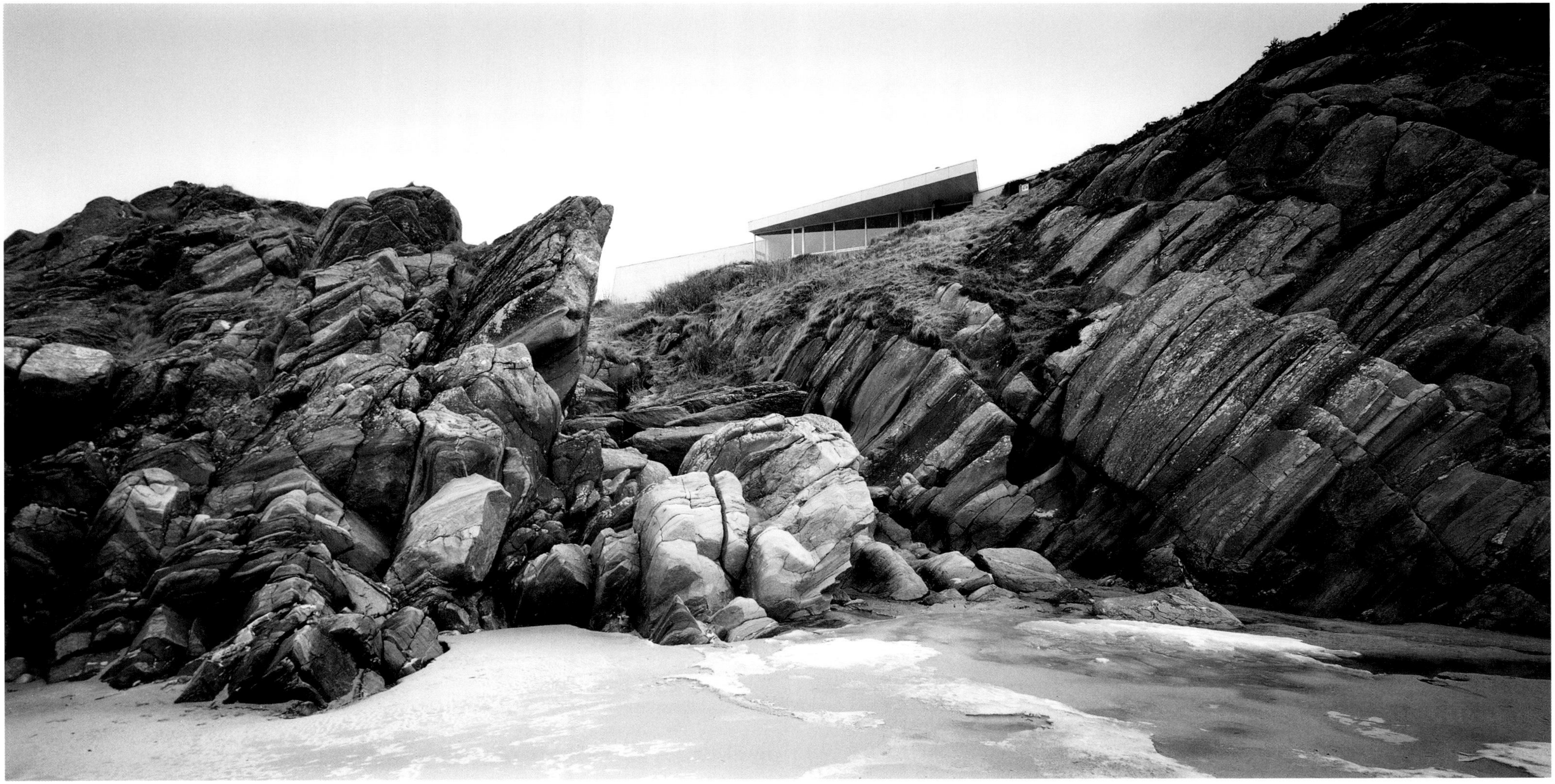

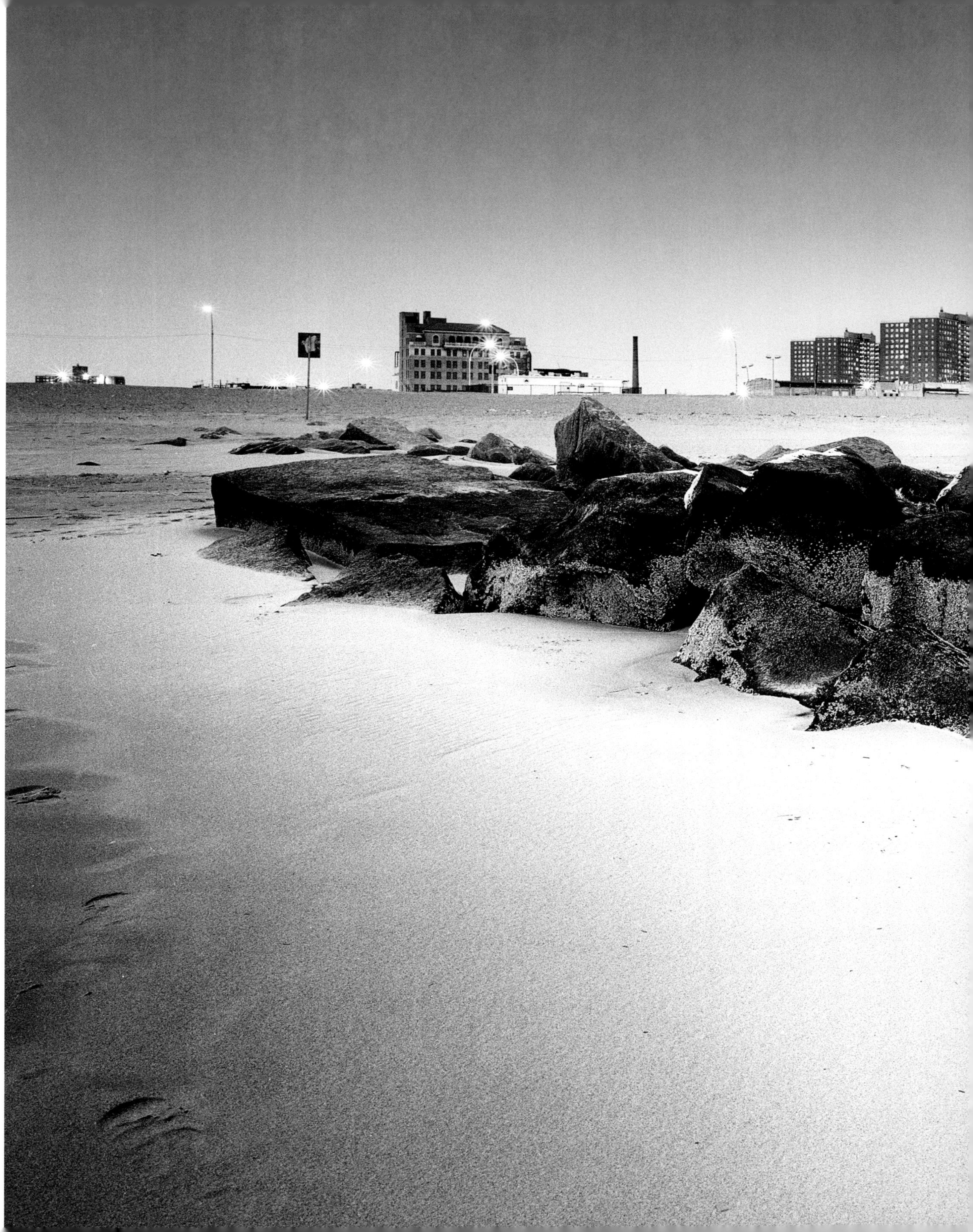

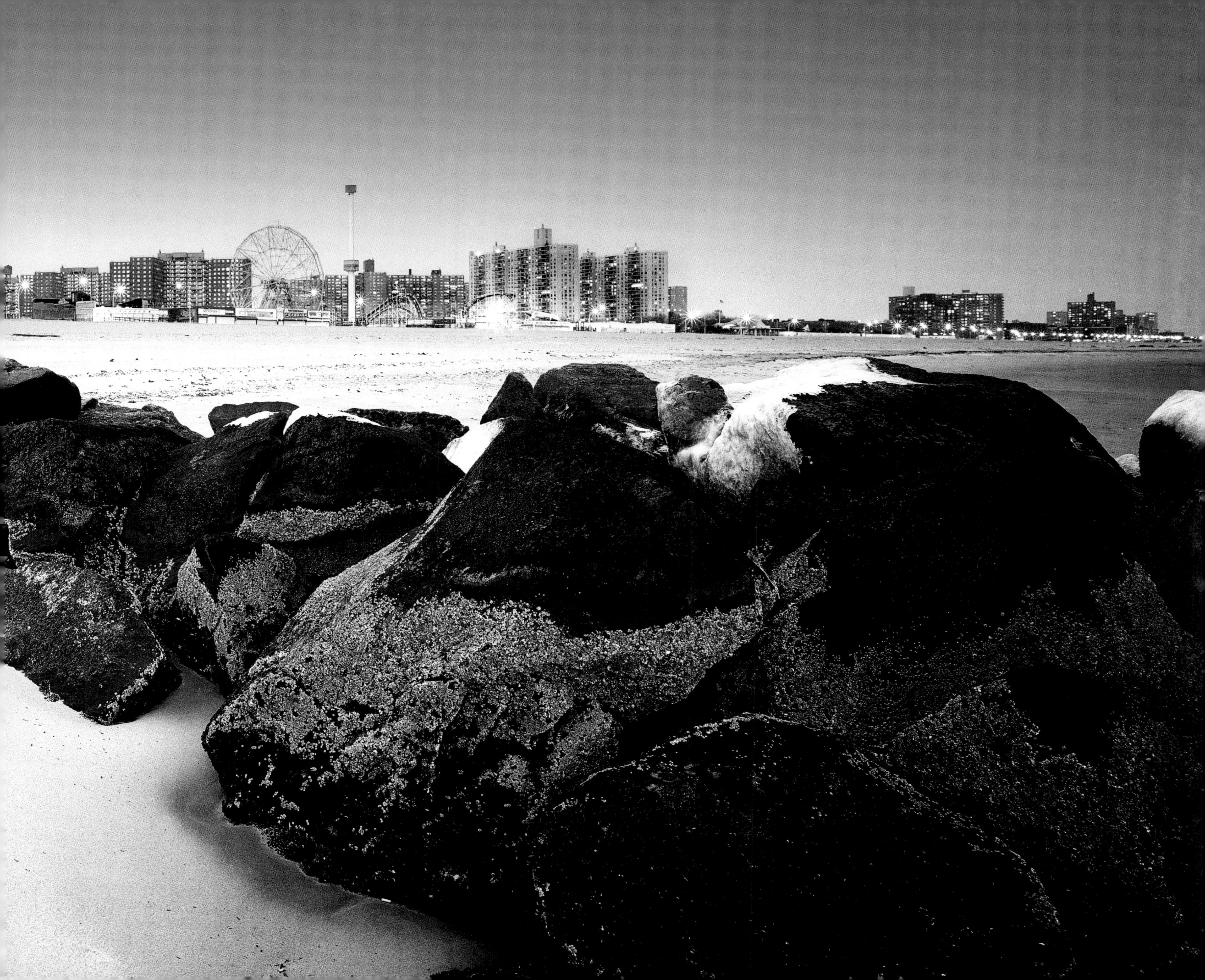

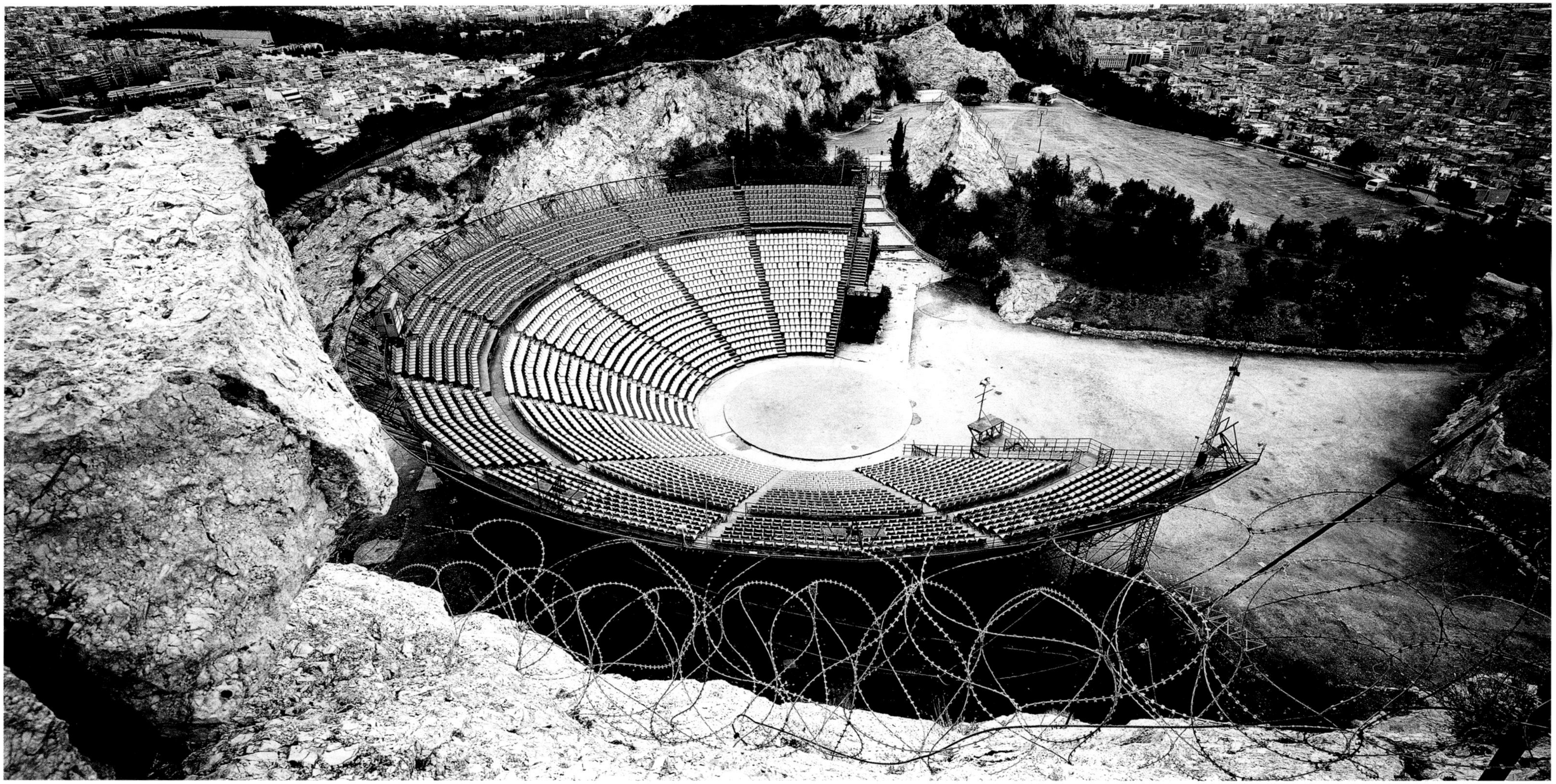

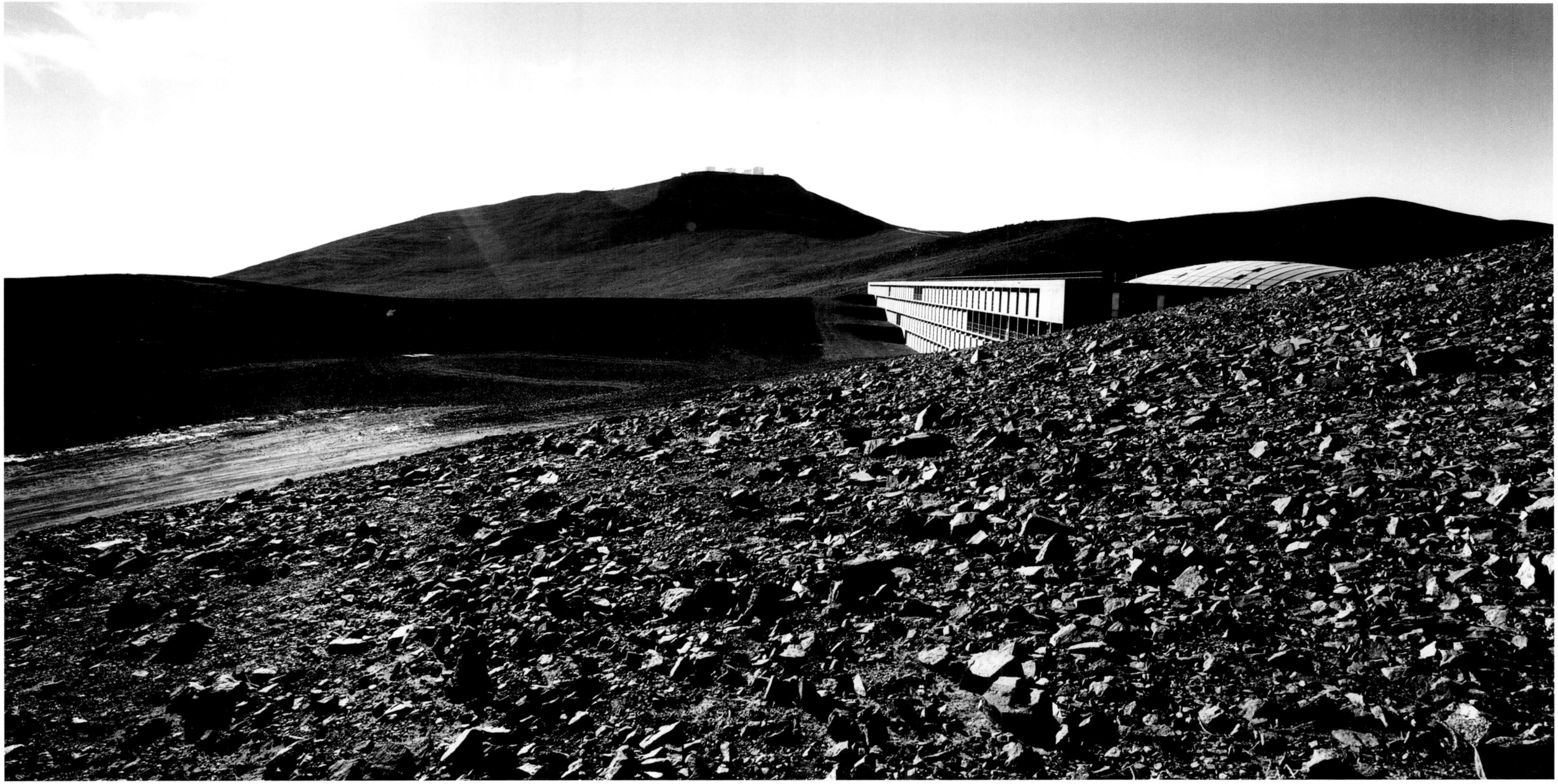

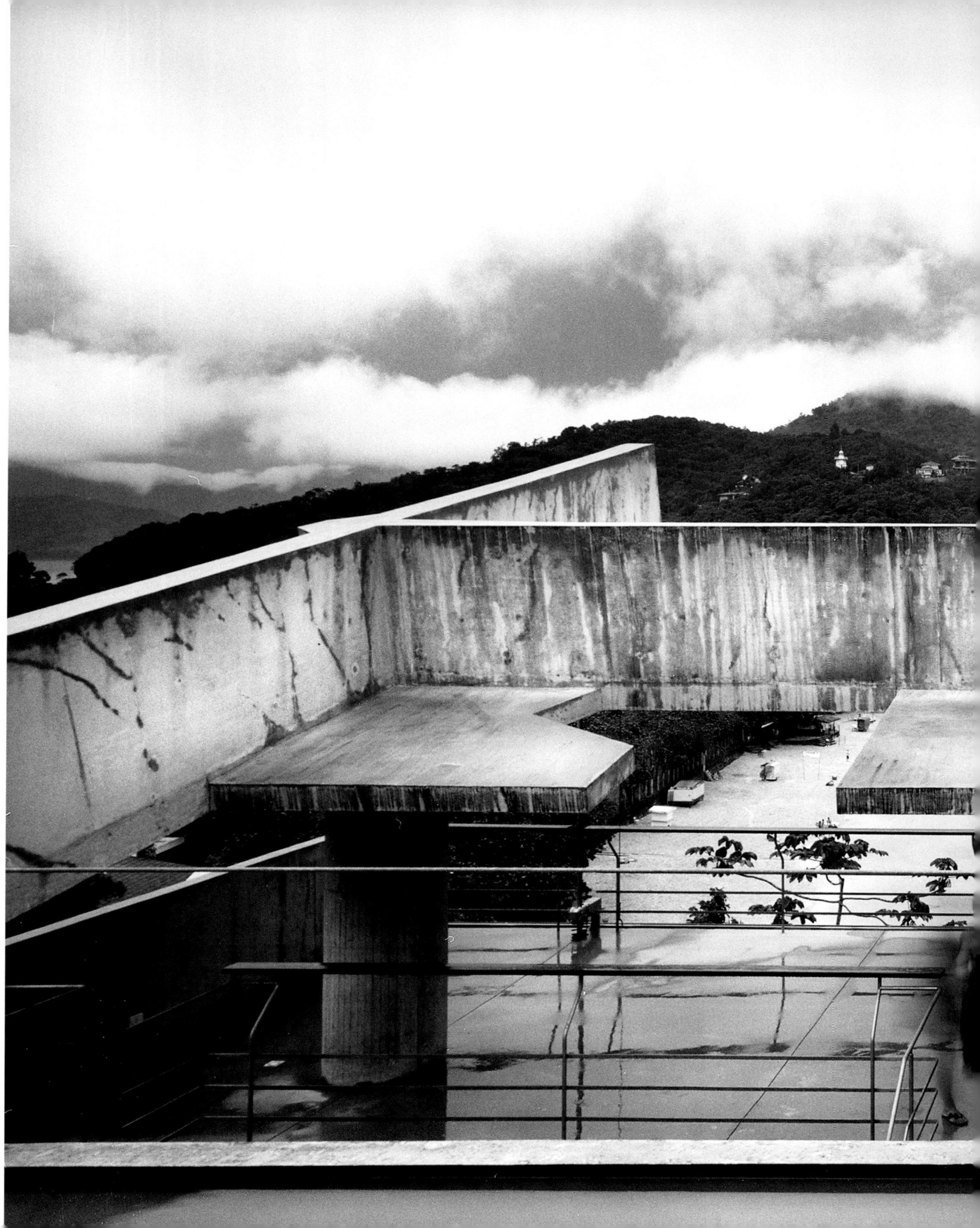

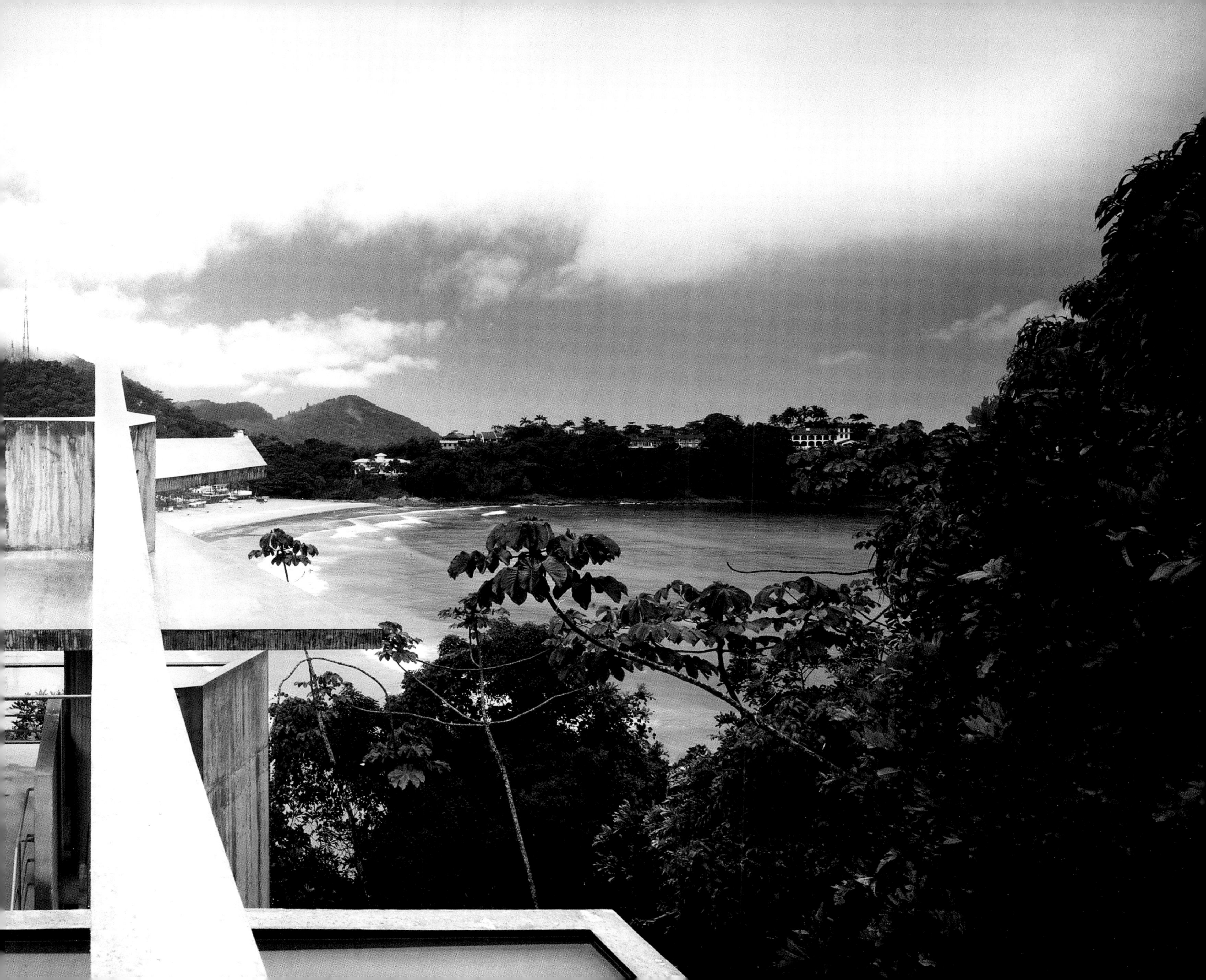

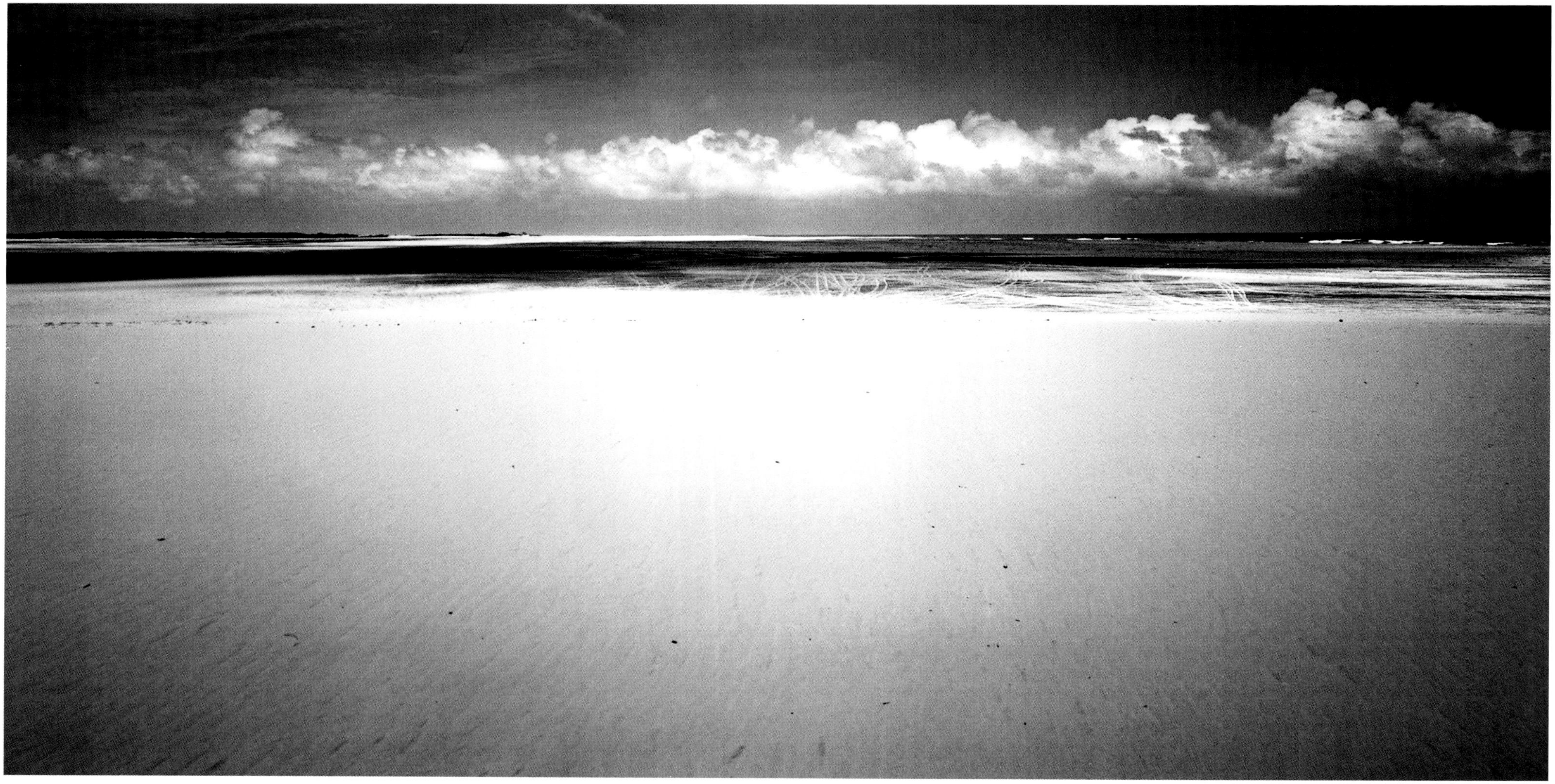

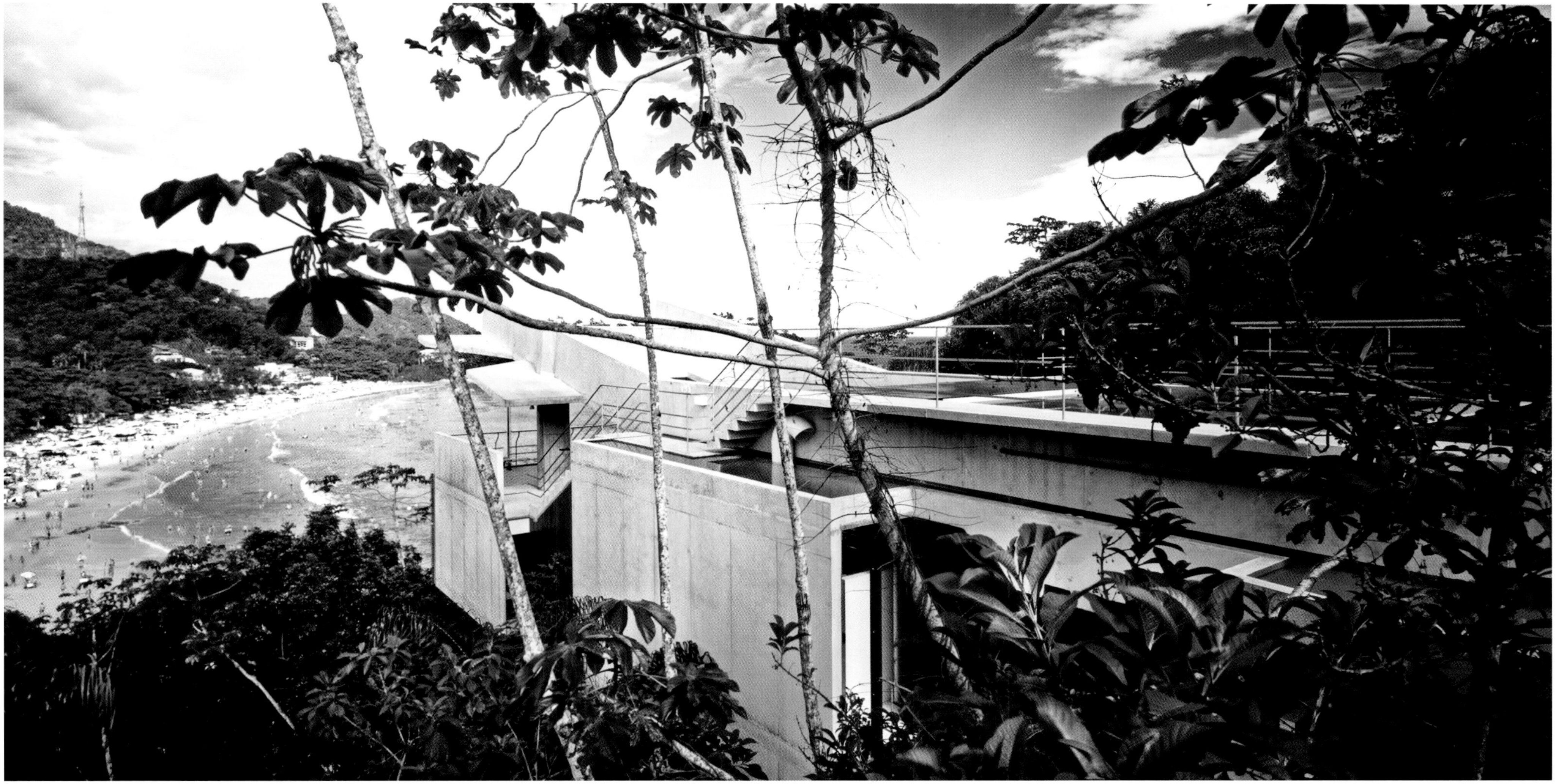

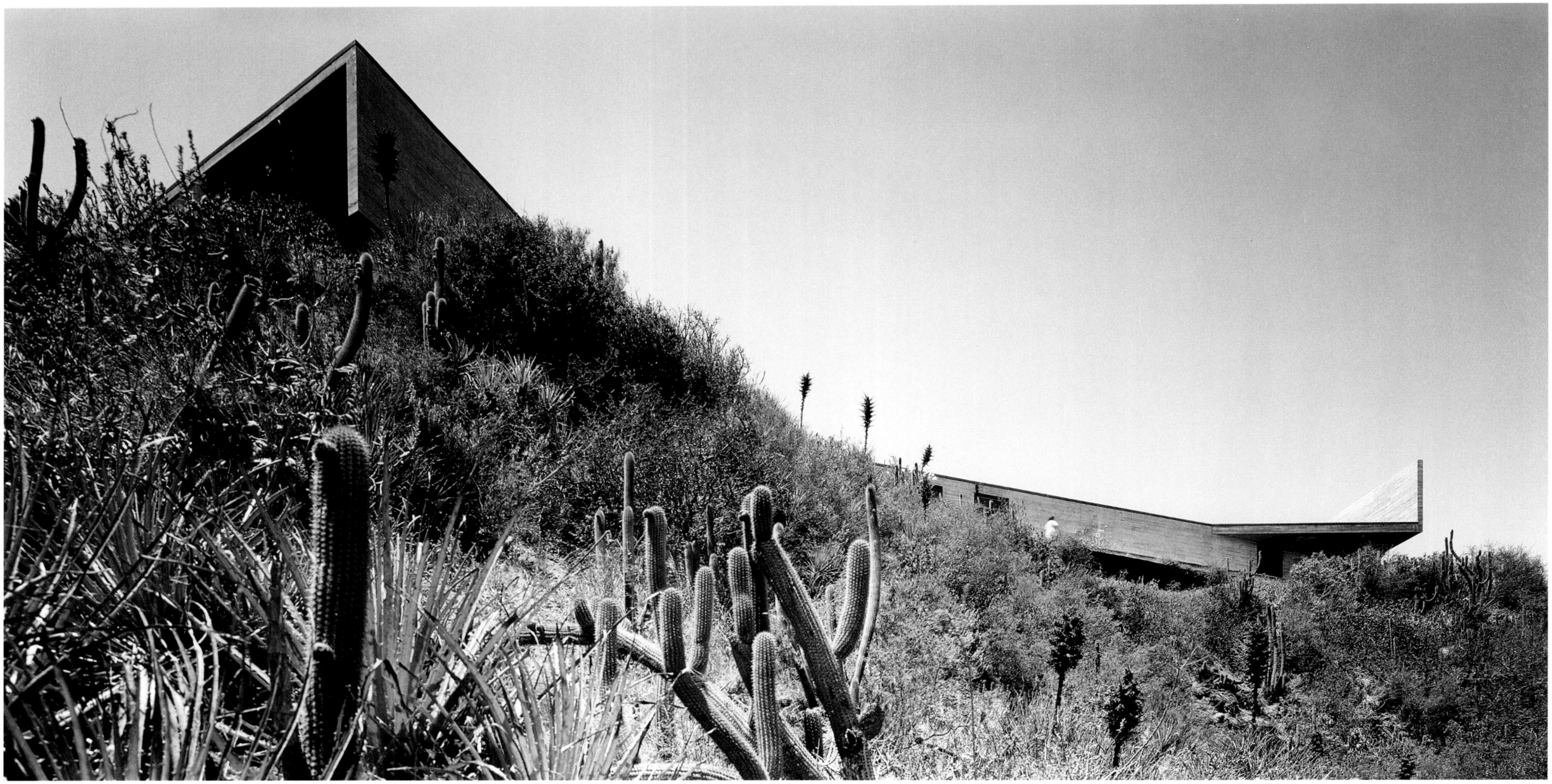

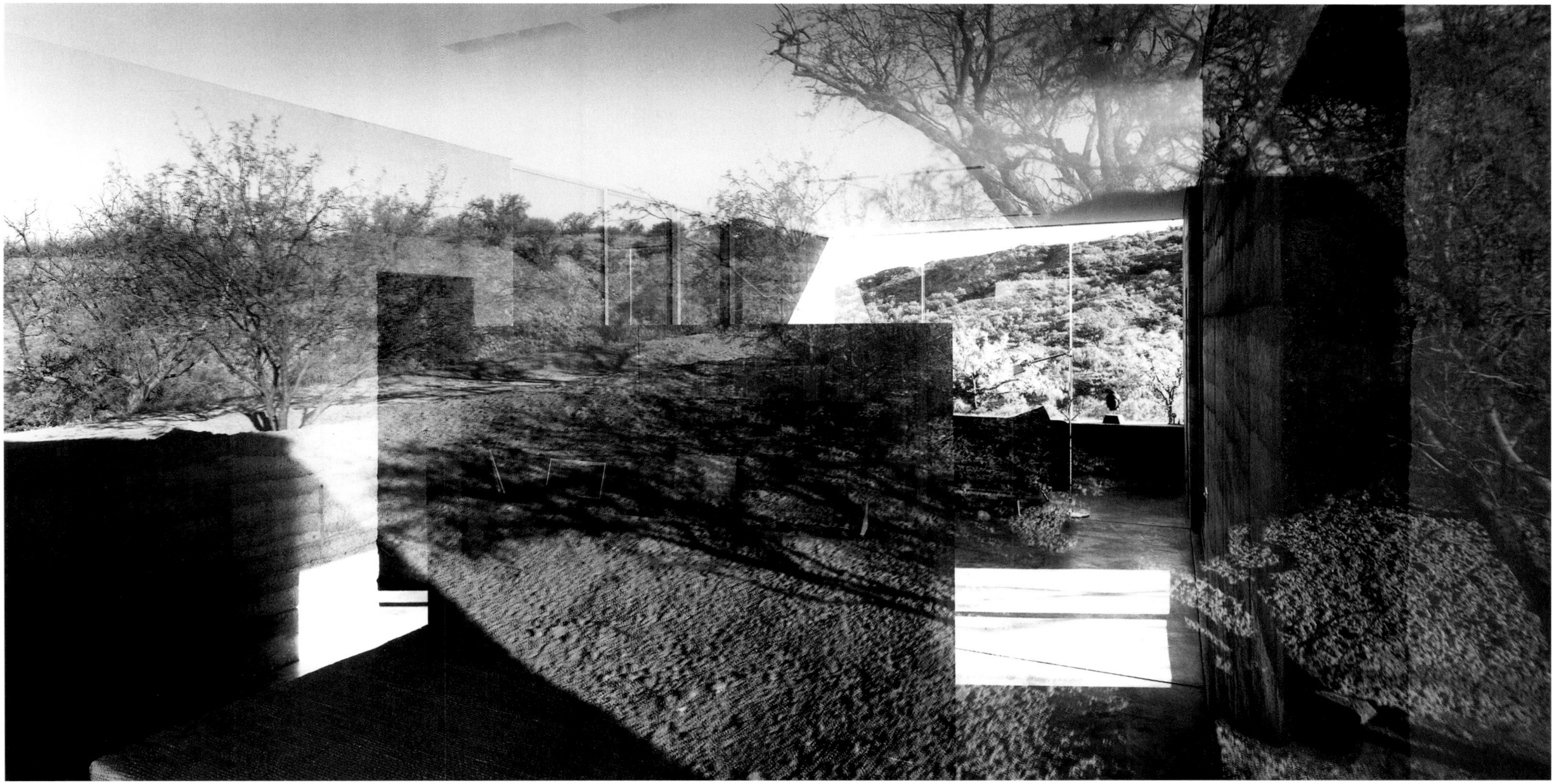

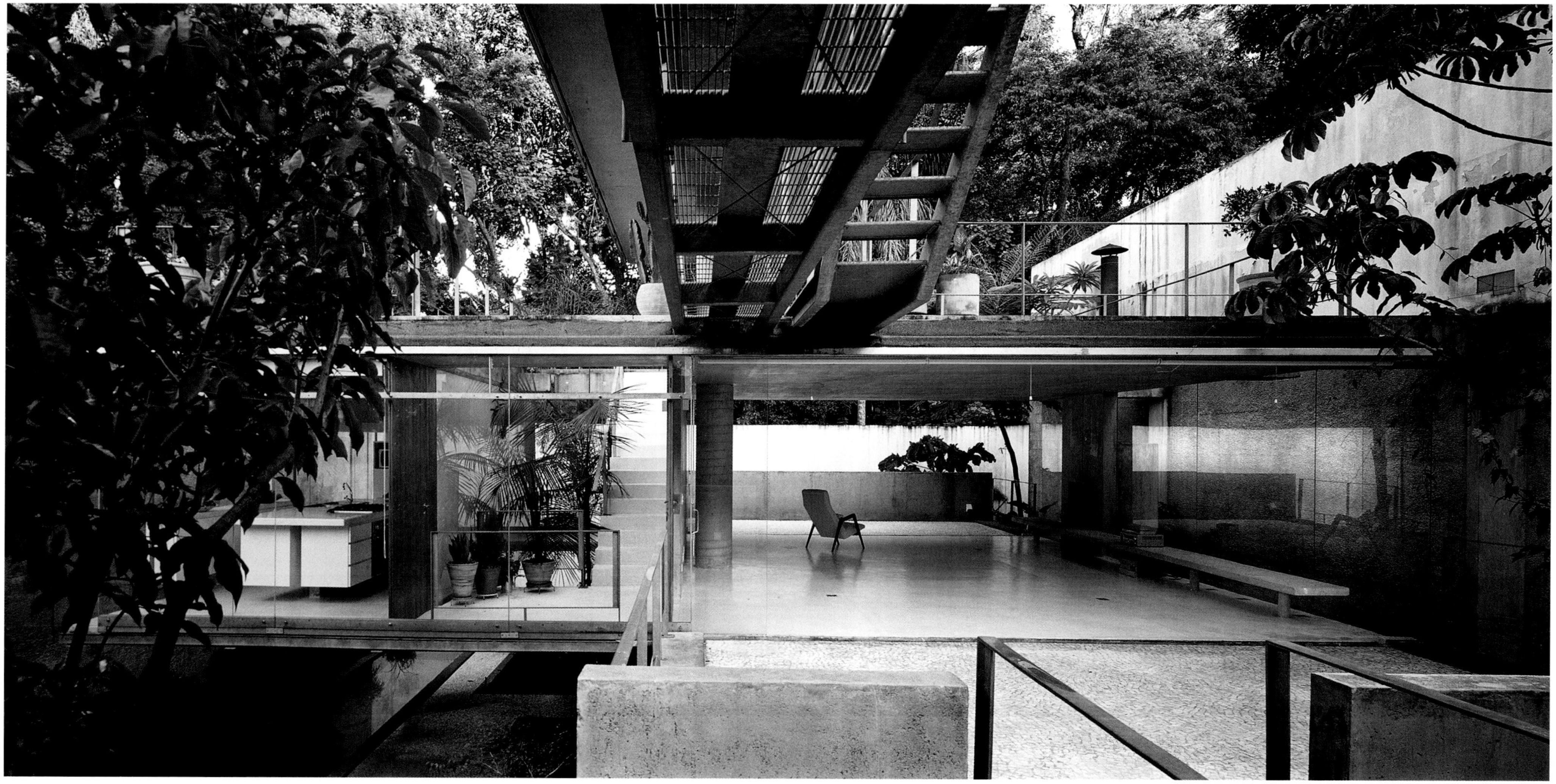

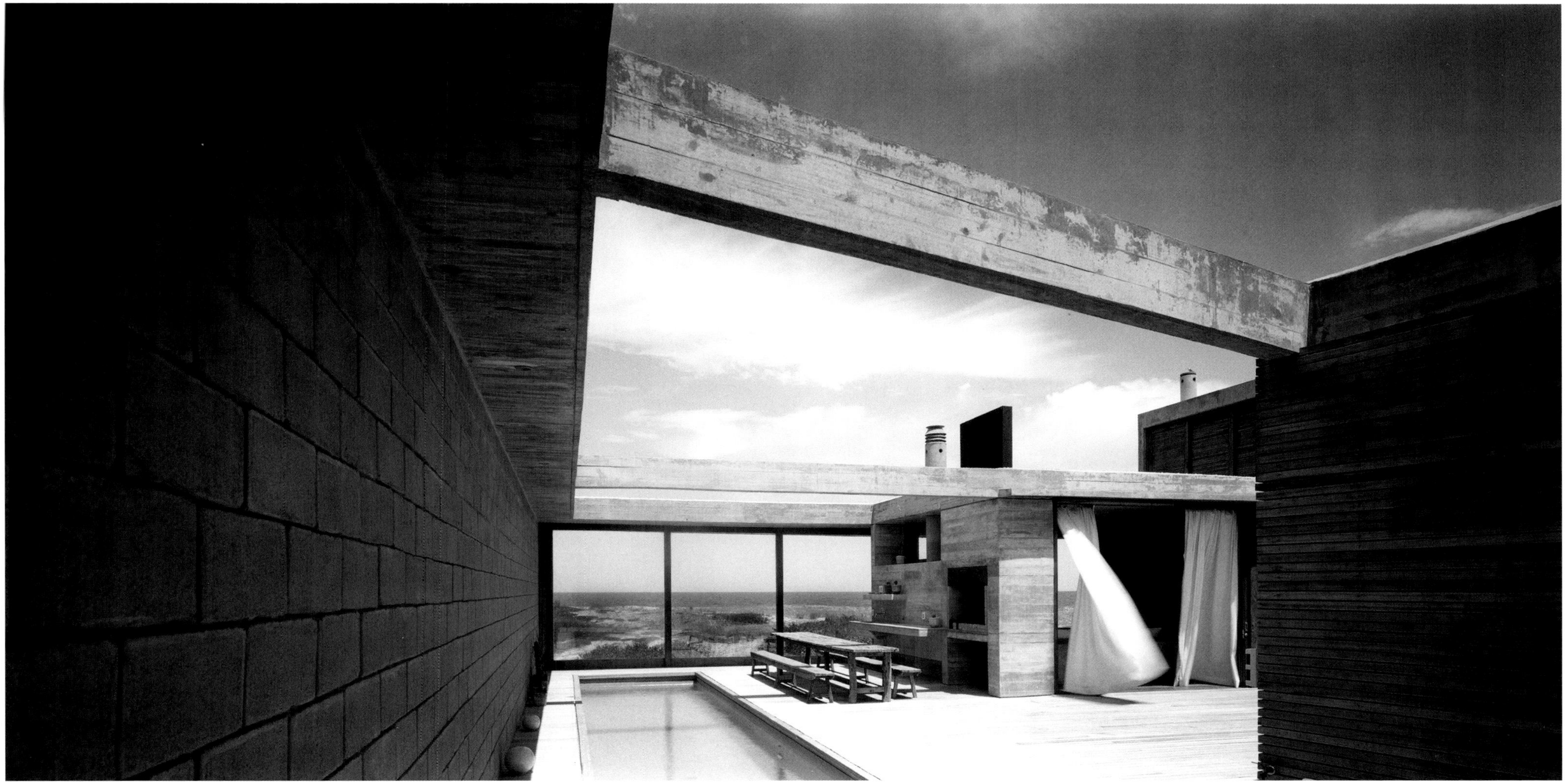

JILLY TRAGANOU

Botanizing Architecture

ERIETA ATTALI BEGAN HER CAREER as an archeological and landscape photographer before venturing into architecture, and in the last eighteen years has compiled an expansive body of work spanning four continents: Europe, Asia and the Americas. In this book rather than provide descriptive representations of specific buildings and landscapes Attali chooses to focus on relations between architecture and landscape. Buildings for Attali rarely seem to function as "machines for living." Rather, these structures are treated as found objects that have surrendered to nature, leading the photographer's gaze to explore and rethink the world. At times it even seems that her chronic preoccupation with architecture has established such deep structures in her photographic approach that even bare landscapes are photographed as if they were buildings themselves. In this book, architecture and landscape are not intellectualized constructions, but deeply sited materialities—earthly substances experienced with a fullness of sensorial presence.

By having chosen to trot around the globe rather than ground herself in a specific place, Erieta Attali has inevitably compiled a body of work that, like herself, recognizes no boundaries. Yet this is the first presentation of her work that boldly acknowledges the insignificance of bounded spatialities. The photographs in this book celebrate the importance of place, by suggesting spatial affiliations that lie beyond physical proximity or socio-cultural stipulations. Unlike previous presentations of her work that were attached to specific national confinements (Greece, Japan, Chile), this book sets free landscapes and architectures to establish new affinities: the Japanese with the Uruguaean, the Swiss with the Chilean, the Greek with the American West. The images of Solano Benitez' father's memorial in Paraguay evoke the sound of water streams, and the same mystical quality of nature as an ancient Japanese cemetery in Kamakura. In a similar way, the dense vegetation of the tropics threatens to overturn the tight geometry of modernist structures, to botanize architecture, to eliminate any national or regional specification for the sake of a deeper quality of the earth that spans continents. As if signaling a post-national imagining of the world, nation-bound characterizations mean very little within the new spectrum of relationships that emerge through Attali's pictures. As if a by-product of an old world order, language insufficiently describes this type of geographical reordering. Whether the architects whose work is featured in this book would accept this suspension of their national or regional attachments for such broader affinities is a moot point. The photographer has the license to reorder the world and to show relations of which the architects or inhabitants may be oblivious or to which, in some cases, they many even be opposed. This does not come as a surprise given her personal trajectory that spans the Mediterranean (Algeria, Israel, Turkey, Greece), and her personal choice to deny the notion of home for what is yet unfamiliar.

Traveling through the pages of this book reminds me of a recent visit to the Brooklyn Botanic Garden conservatory where botanical classifications are based on climatic zones: the temperate, the tropical, the desert. Climatic zones span the continents suggesting a new way of classifying places in the world through analogies of climate and flora. Here, despite the physical distances separating them, the Sahara, Andalusia and the American West,are shown to be alike. Similarly, the jungles of South America form a conglomeration with those of Asia or Polynesia. It is the tropics and the desert that seem to most interest Attali,or what she calls "remote landscapes." Attali even treats these two types of landscapes as if they belong to the same category: landscapes with not only formal analogies, but terrains in which habitation, as every other human act, including her own act of photography, is being challenged. Here architecture, despite its geometry and rationality, seems nothing more than a product of re-ordered materiality, less powerful than the simple mound that one discovers while walking in the woods that Adolf Loos saw as the predecessor of architecture. These deserted landscapes point more towards the unbuilt, the uninhabited; in them architecture seems almost an incidental, discontinuous gesture.

But it is not the search for the exotic, the challenge of breaking the frontier, or the imperial gaze of the colonizer that draw Attali to these locations. The tropics for instance are reminiscent of her childhood in the Princes Islands in the Bosphorus of Istanbul. The desert is familiar from her exposure to Mediterranean landscapes in the summer, when, as the historian Fernand Braudel has observed, the entire sea basin is being invaded by the hot dry air of the Sahara. Attali's photographs not only defy traditional geographical categorizations but also evoke a post-archeological view of the world that obscures the temporal specification. Her pictures are almost undated; they focus neither on the present nor in the past. Unlike the work of many of her contemporary traveler-photographers, they are no prophecies for the future.

As you look at Attali's pictures, you travel through a reordered planetary topography that awakens the senses into their highest capacity. Here the photographic lens augments the eye by offering overexposures and panoramic views that the naked eye could not grasp. Through it, you dare to look straight to the sun, when otherwise the eyes would hurt. As a result a new sense of coloration and materiality is evoked, such as in the picture of Aurelio Galfeti's house in the Aegean where the white buildings of the Aegean islands appear to be black starkly contrasting with the whiteness of the horizontal concrete plate mirroring the reflectiveness of the sea. The plate of crushed marble breaks the slickness of the sea, bringing memories of the Atacama desert, and echoing the texture of the surrounding hills that we have seen in pictures of Deca's architecture on the nearby island of Antiparos. But in addition to these seeming optical illusions, a product of her mastery of the photographic lens, the breathtaking quality of Attali's visual manifestations evokes a strong sense of tactility, accentuating the textures and the materiality of the world: from masses of extreme softness (vapor, water, or even stone illuminated by the sun) to a roughness that poses a threat to the body.

What kind of spectatorship does this work bring forward? What does it tell us for our current age? Attali's work seems to be apathetic to the virtual aspects of the world that govern our daily reality. Her photographs are not seduced by Google Earth views of the world. Her work makes no commentary on the commodification of visual representations that surround us, nor does the digital seem to concern her. Her work is deeply embedded in the sensuality of the body: it involves long walks, early rising, carrying heavy equipment. Her craft relies on approaching the tangible world with an uncompromising persistence. Still, the process that makes all this travel possible depends heavily on information technologies. Attali spends no day without her laptop, her dates of travel are determined by persistent attention to satellite weather forecasts; her photographs are becoming possible by painstaking networking with architects, clients and various impresarios. She travels trans-continentally several times a month and has no home. She draws the most out of the virtual in order to reach the apogee of experience in the physical world; reminding us that the earth is still there.

Fernard Braudel, *Memory and the Mediterranean,* (New York: Alfred A. Knopf, 2001) 8.

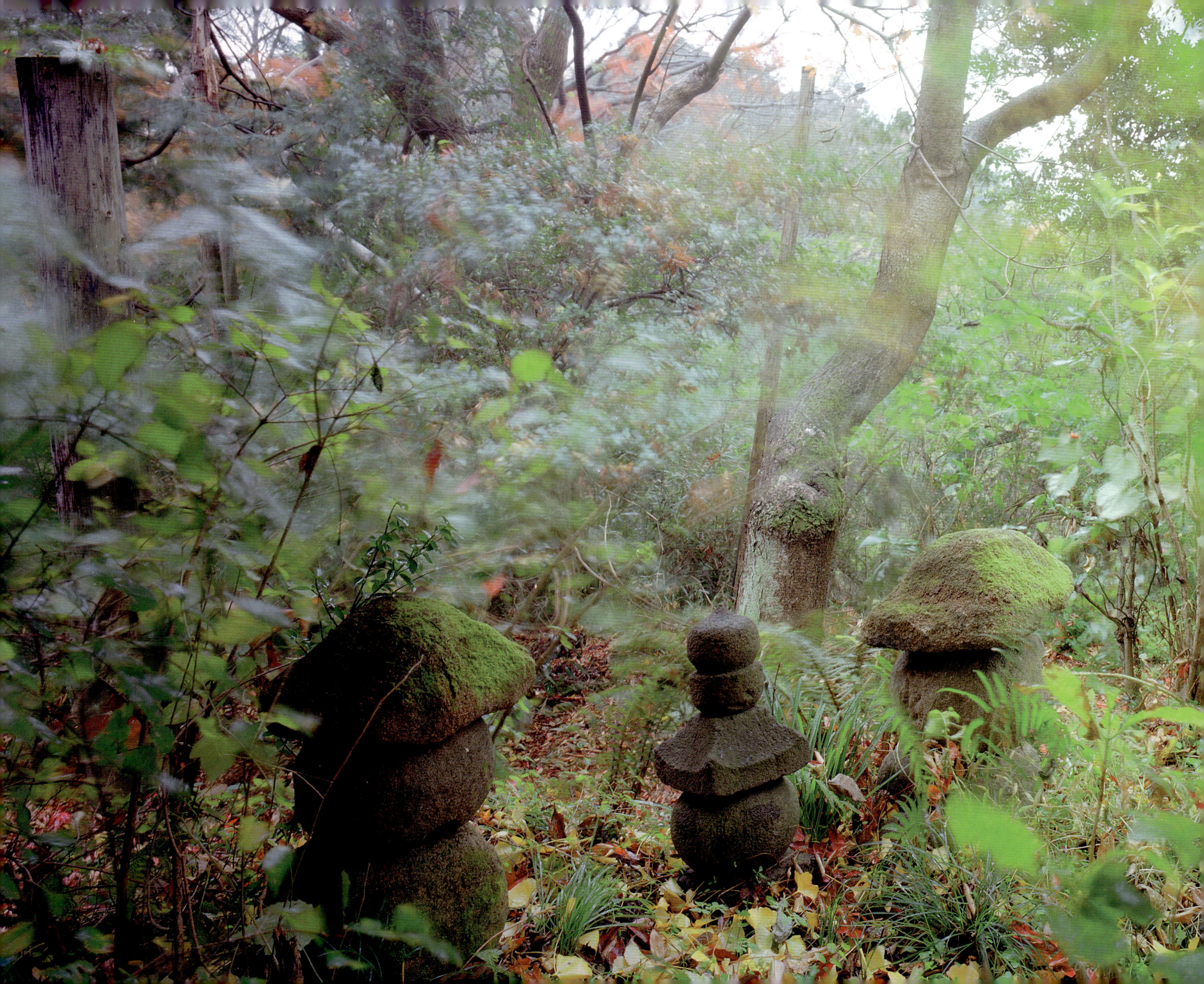

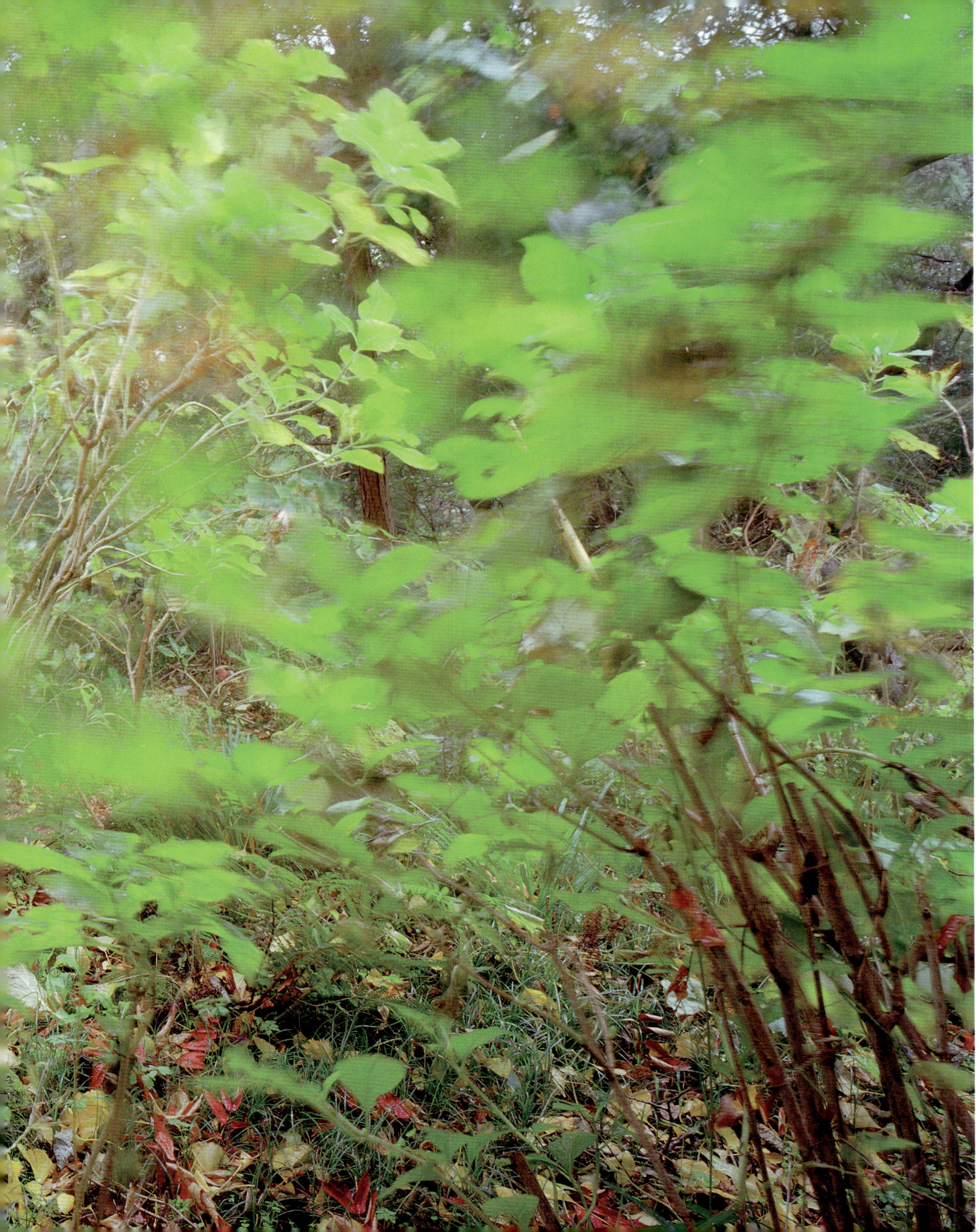

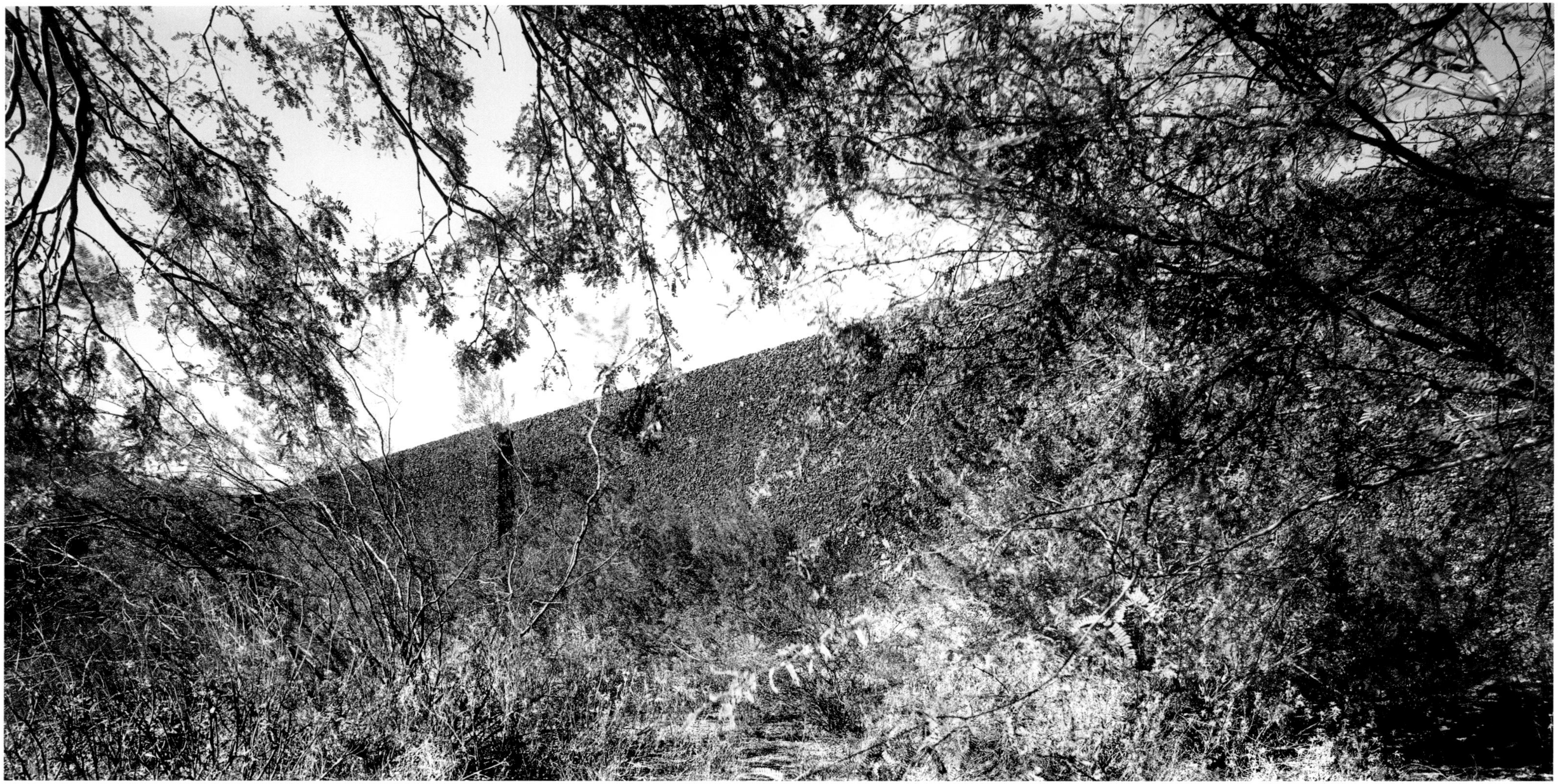

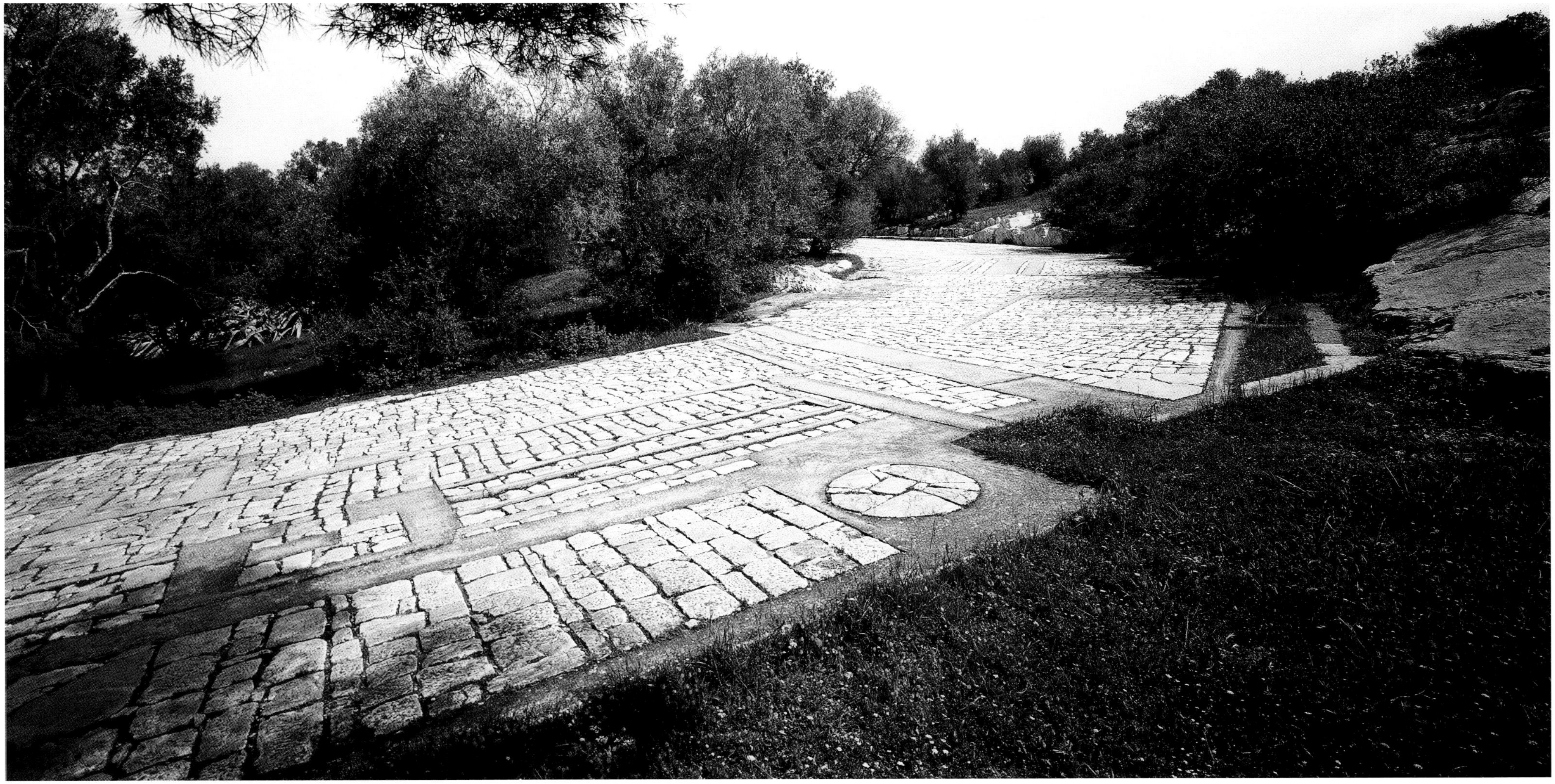

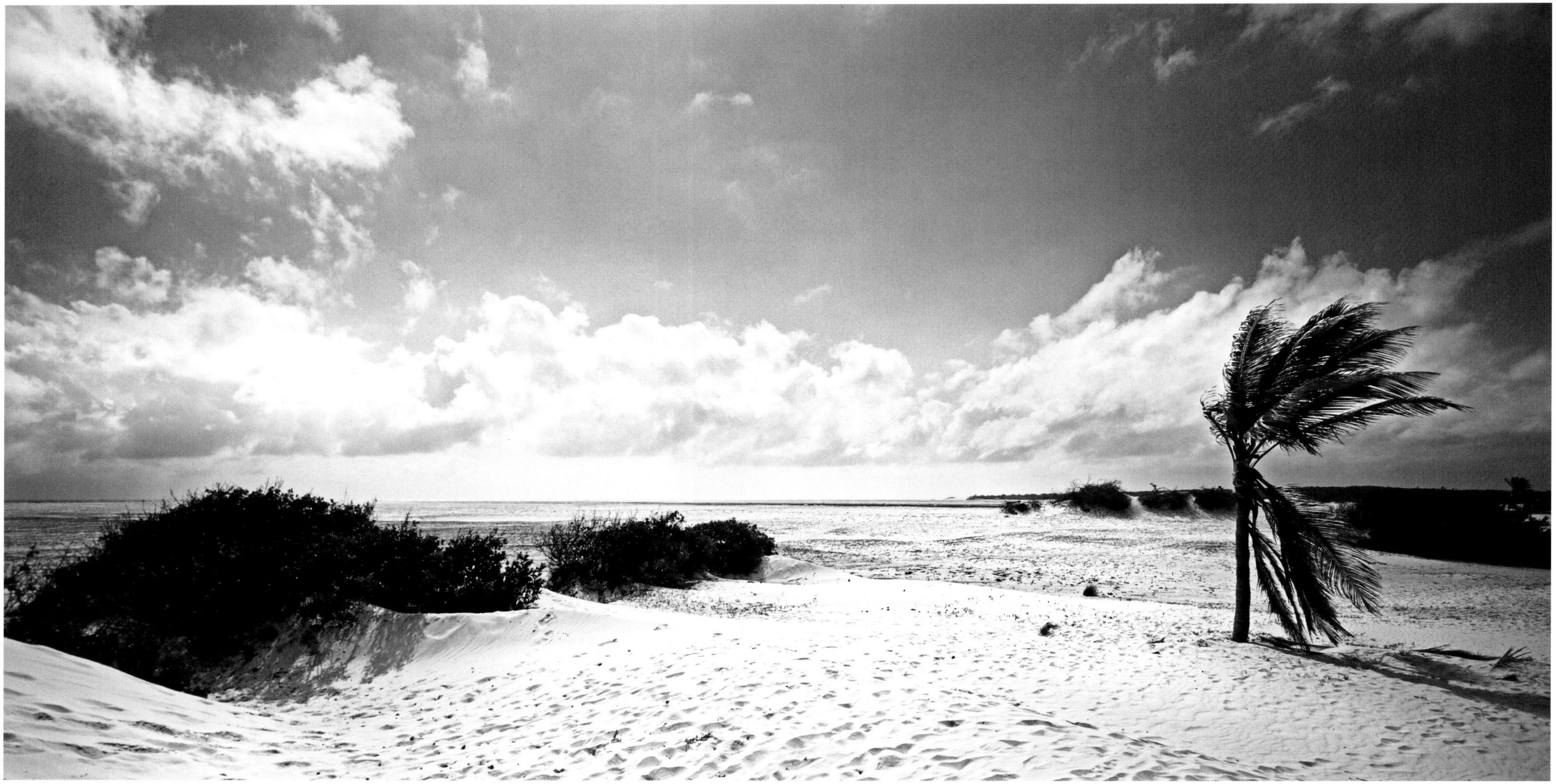

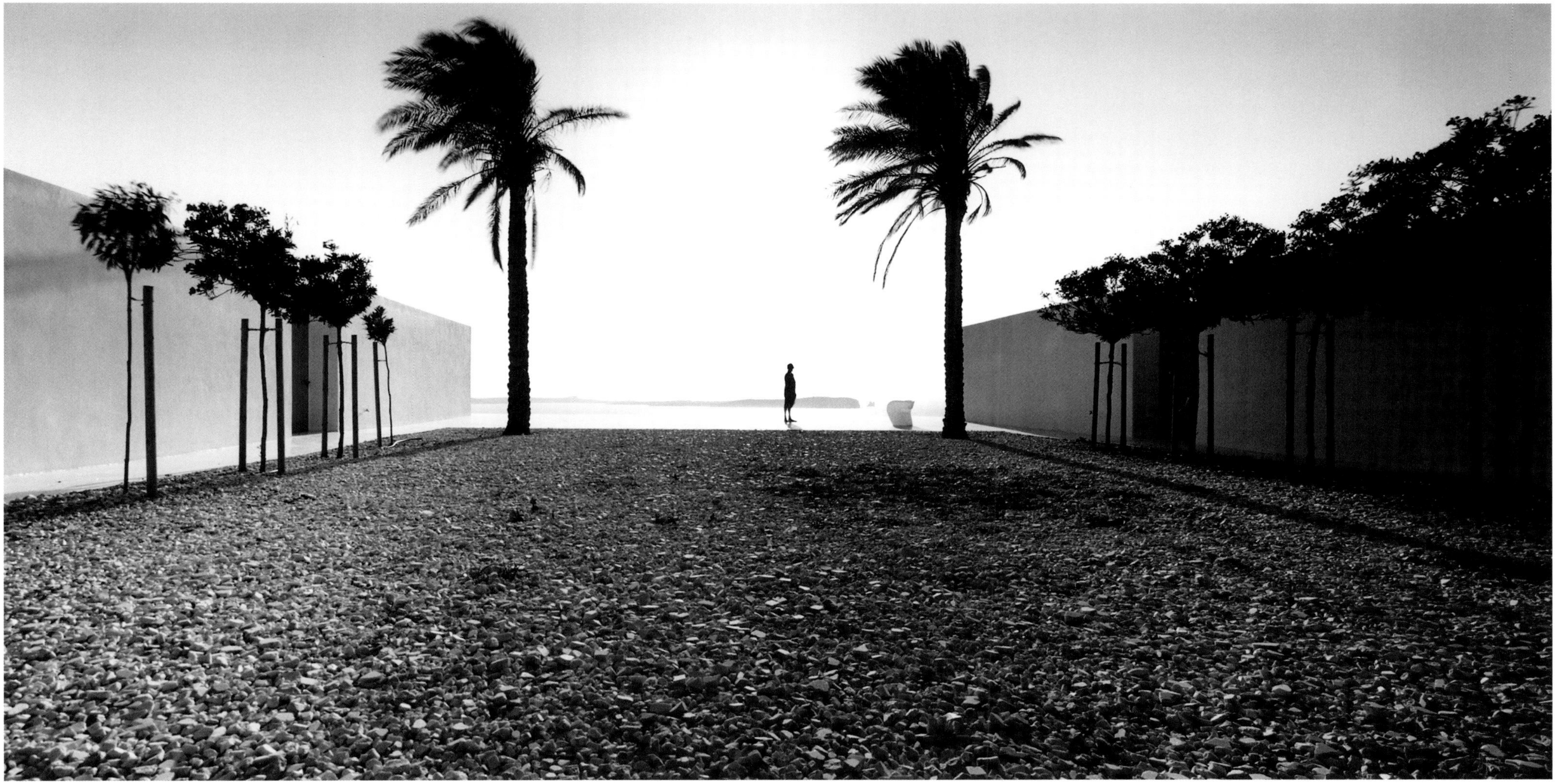

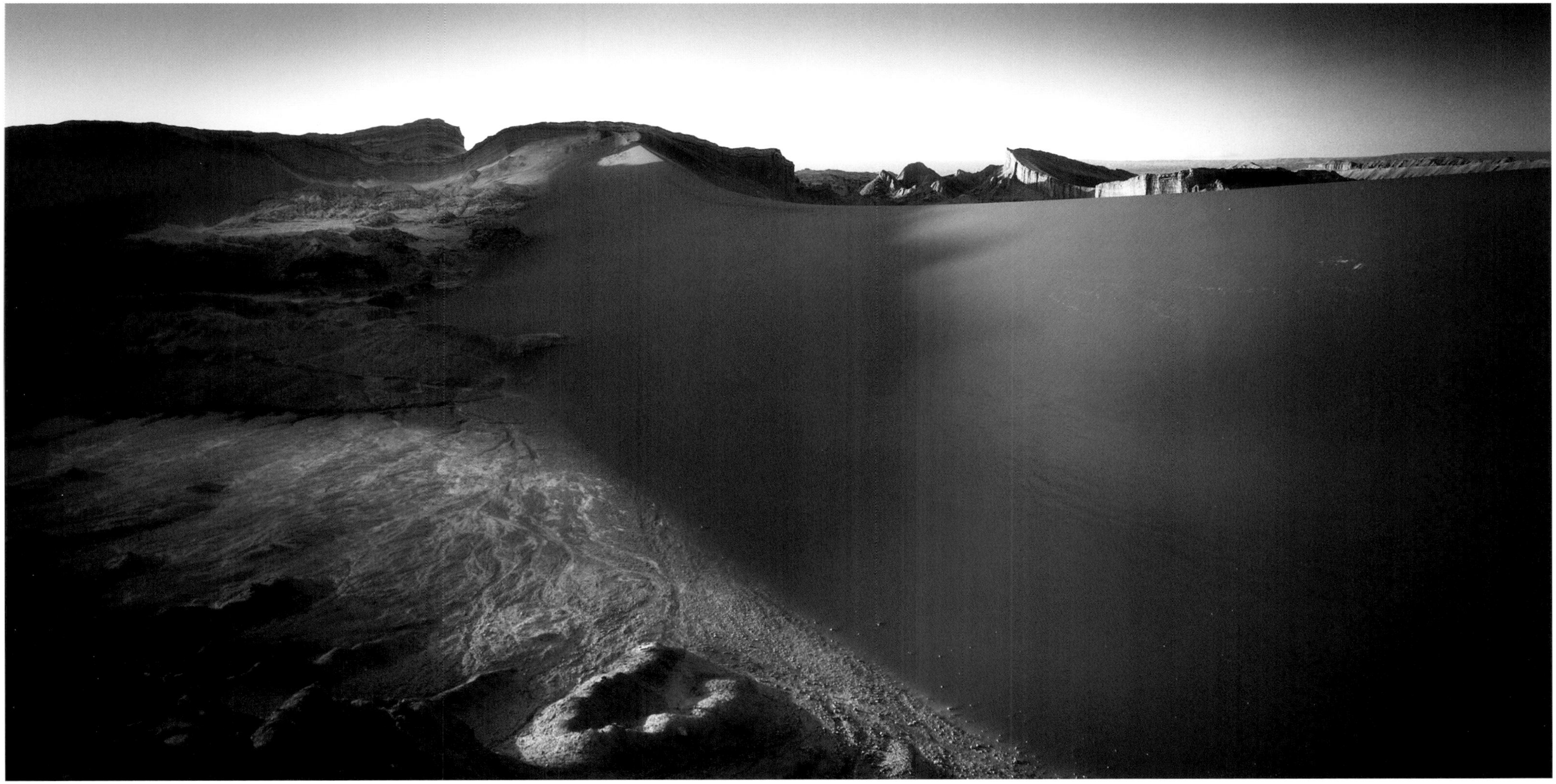

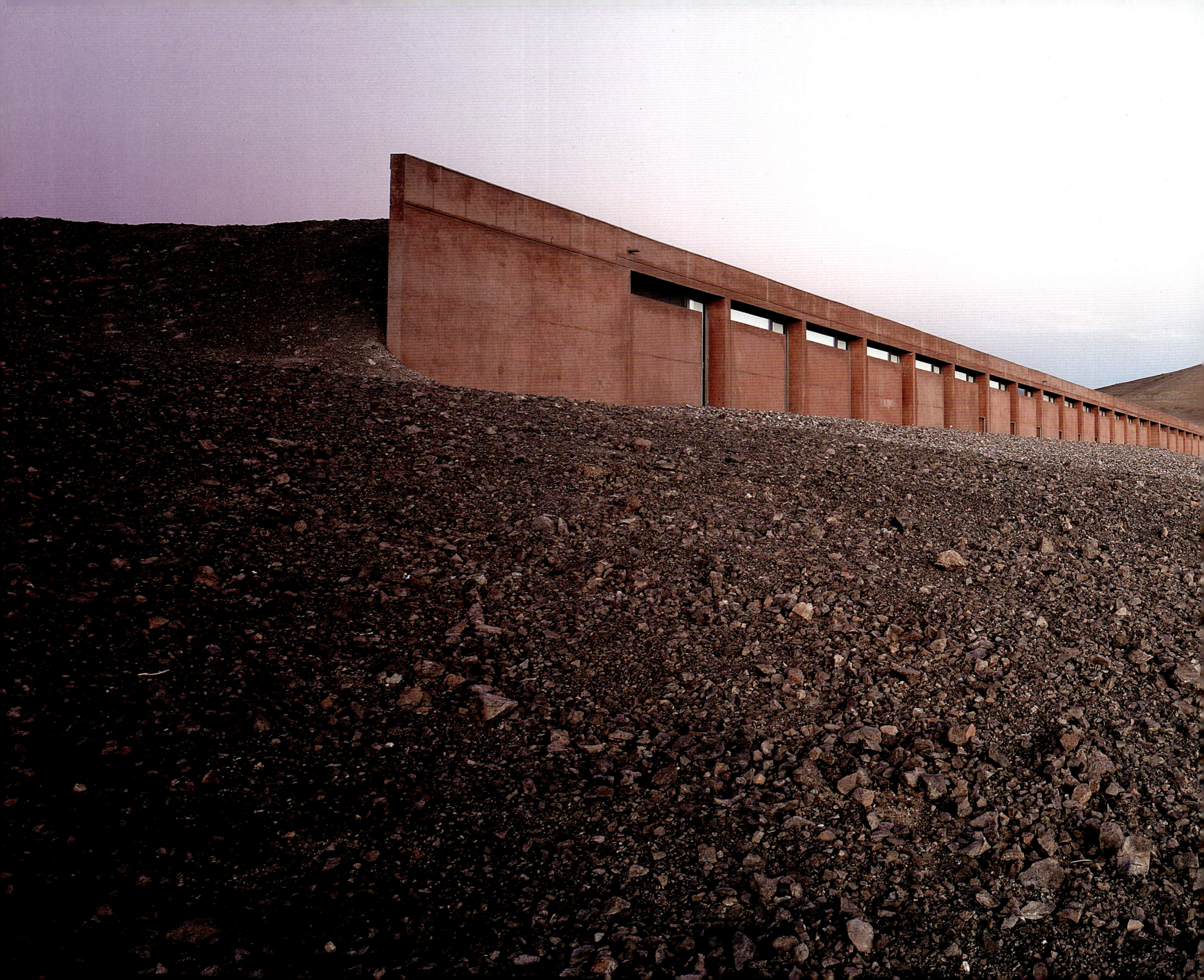

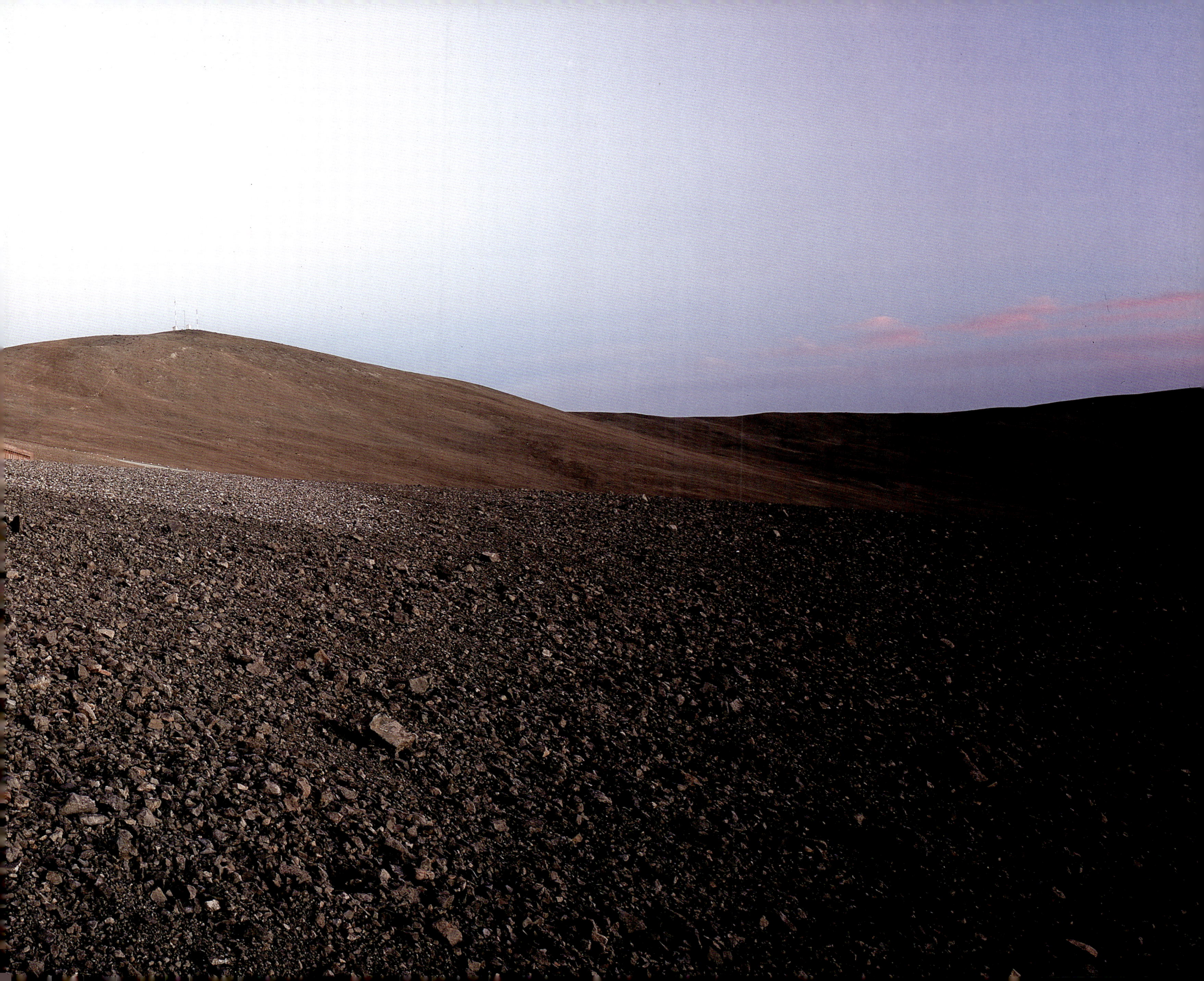

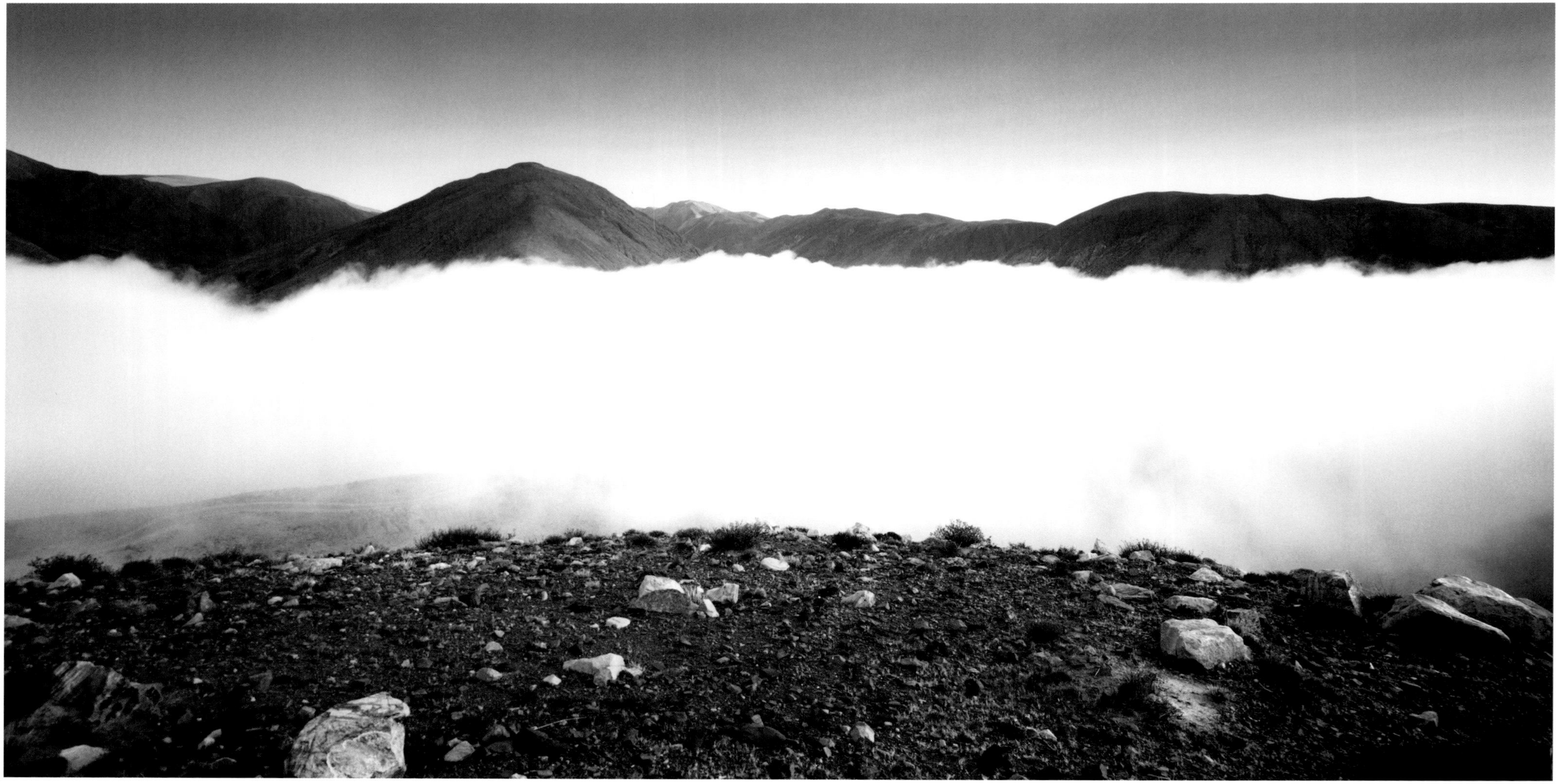

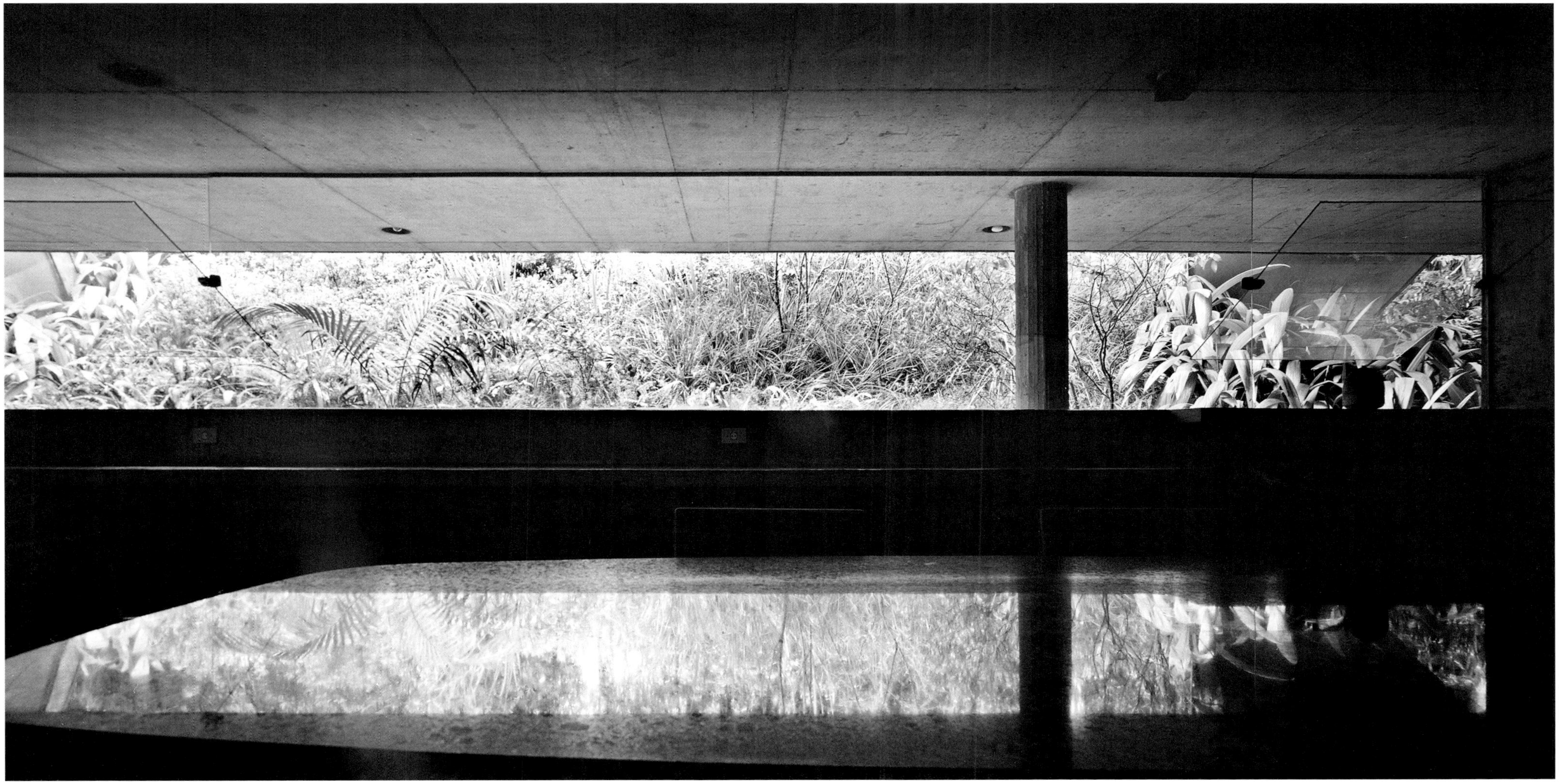

ALESSIO ASSONITIS

Consecrated Spaces, Sacred Journeys

"Do you want to renounce your faith in poverty and build grand palaces?"

WITH THIS STERN ADMONITION, Dominic de Guzmán (1170–1221) vehemently reprimanded a friar at the convent of San Niccoló delle Vigne in Bologna for having erected walls slightly taller than what was prescribed by the Order's unwritten rules. At this very early stage in the history of the Dominican Order, the meter of aesthetic pauperism was determined more by appearance than by regulation. The founder himself asserted that friars should live in small edifices and poor cells, suitable solely for study and sleep. For a brief time only, construction tempered by a genuine espousal of *paupertas* (poverty) served as the rule. These vague directives, however, were bound to be short-lived: an explicit pandect soon became necessary. The first Dominican Constitutions instructed Friar Preachers to live in *mediocres* (modest) and *humiles* (low) spaces, defined by precise measurements, in buildings that would not be a cause for scandal. Moreover, their churches had to be void of *superfluitates* (excesses) and *curiositates* (distractions). Yet, even these regulations turned out to be insufficient. Dominican conventual architecture was marred by infractions and subsequent emendations to the Constitutions. In all effects, paupertas was relinquished in favor of practicality and comfort, not to speak of frivol— particularly throughout the Italian Renaissance—sporadic though uncompromising pauperistic movements reacted against this laxity, but without success.

There are a number of important elements that trace certain converging trends in pauperistic aesthetics and Erieta Attali's visual research. Many buildings featured in her photographs incarnate a strong disdain for "excess" and "distraction," a lack of ostentatious *majestas,* and a conscious neglect of worldliness. Eclecticism and mannerisms are, for the most part, kept to a minimum in favor of a contemporary, often anodyne, stylistic sobriety. Even the use (and reuse) of materials points to a modern reinterpretation of Mendicant aesthetics. Originally bricks, thatch, plaster, and wood were preferred to marble and sandstone, which were considered to be rich materials. Furthermore, the relation of buildings to their respective environment constitutes no "cause of scandal." Some interact with the landscape, in an attempt to blend in, modestly and silently. A few others stand apart, emerging monumentally from nature's viscera. Attali investigates the interaction of these structures with their surroundings, challenging the viewer to detect the patterns of interpenetration between landscape and building. Even the most declamatory architecture is ensnared by the precepts of Attali's distinctive aesthetic tenet. As the viewer perceives the relationships between the constituent parts of the photographic composition, elements of visual dissent seem to be tamed by the unifying harness of tonal and atmospheric synthesis. Even nature itself is subject to this process of homologation, which is intrinsic to Attali's creative process.

Upon closer examination, there is a strong sacral component in this selection of images. Attali extracts hermitages, pilgrimage trajectories, and sacred precincts from contemporary structures that have marked their respective settings, often in a violently dialectical manner. She seems to begin with a palimpsest already indelibly inscribed with pagan and monastic architectural lexicons when composing her visual compositions. Private homes become isolated sanctuaries and monasteries in austere landscapes. These structures appear to be accessible to few while remaining visible to many. Some serve as refuges from contemporaneity; some function as lighthouses guiding the wayfarer to his destination. This latter theme is explored by Attali on a number of occasions, each time taking on the guise of a different ilk of wanderer. Footsteps on Maranhão sand dunes in Brazil not only establish a record of a journey, but they also sanction human intervention in the landscape. Similarly, a road plowing through the desolate territories in the Argentine province of Jujuy evokes the harsh travel of medieval pilgrims on the Via Francigena. Attali transposes the theme of the religious itinerary to an urban context. The pathways paved with quarry detritus and ancient spolia, assembled by Dimitris Pikionis in the 1950s, inevitably point to the religious experience of Panathenaic processions. A similar kind of ceremonial function is enacted, perhaps even on a daily basis, on the sloping brick progression of Villa Elisa, rising towards the skies like a Mesopotamian ziggurat. When we finally enter these sanctuaries, we are drawn into their liturgical spaces. Here the surgical precision with which Attali constructs her compositions is immediately perceptible. The exterior court preceding the perspectival *fuga* towards sea and sky in the Galfetti House overlooking the Aegean and the interior of Bernard Tschumi's New Acropolis Museum, punctuated by sculpture from the west pediment, are just two examples of basilica naves carved out from private and public spaces. In both cases, the sacred precincts and their foci—respectively, the horizon line and the Parthenon—are clearly delineated for the performance of secular rituals. Furtively, almost to validate the abovementioned profane areas of congregation, an authentic nave in the Cemetery Hall in Zug is included. A reflection on the glass partition and vertical strips of curtain, reminiscent of the colonnade in a Paleochristian basilica, partly disguise the cross.

Erieta Attali's formal and iconographical solutions—including her attempts to reconfigure the sublime—are subject to her distinctive aesthetic imprimatur. Through this filter, even the most disparate geomorphological and climatic sites appear to originate from the same general location. Not even architecture is spared by this overarching aegis. This operation, however, is performed with utter humility and simplicity, without recourse to shortcuts or curious distortions. Each image is the product of careful research, an incessant eradication of the superfluous and affected. Her personal *paupertas* is found in the revelation of the intimacy of sacred spaces and in the preservation of the deafening silence of waterfalls, deserts, and oceans. At all times, she remains devoted to this rule, and is never tempted to *build grand palaces.*

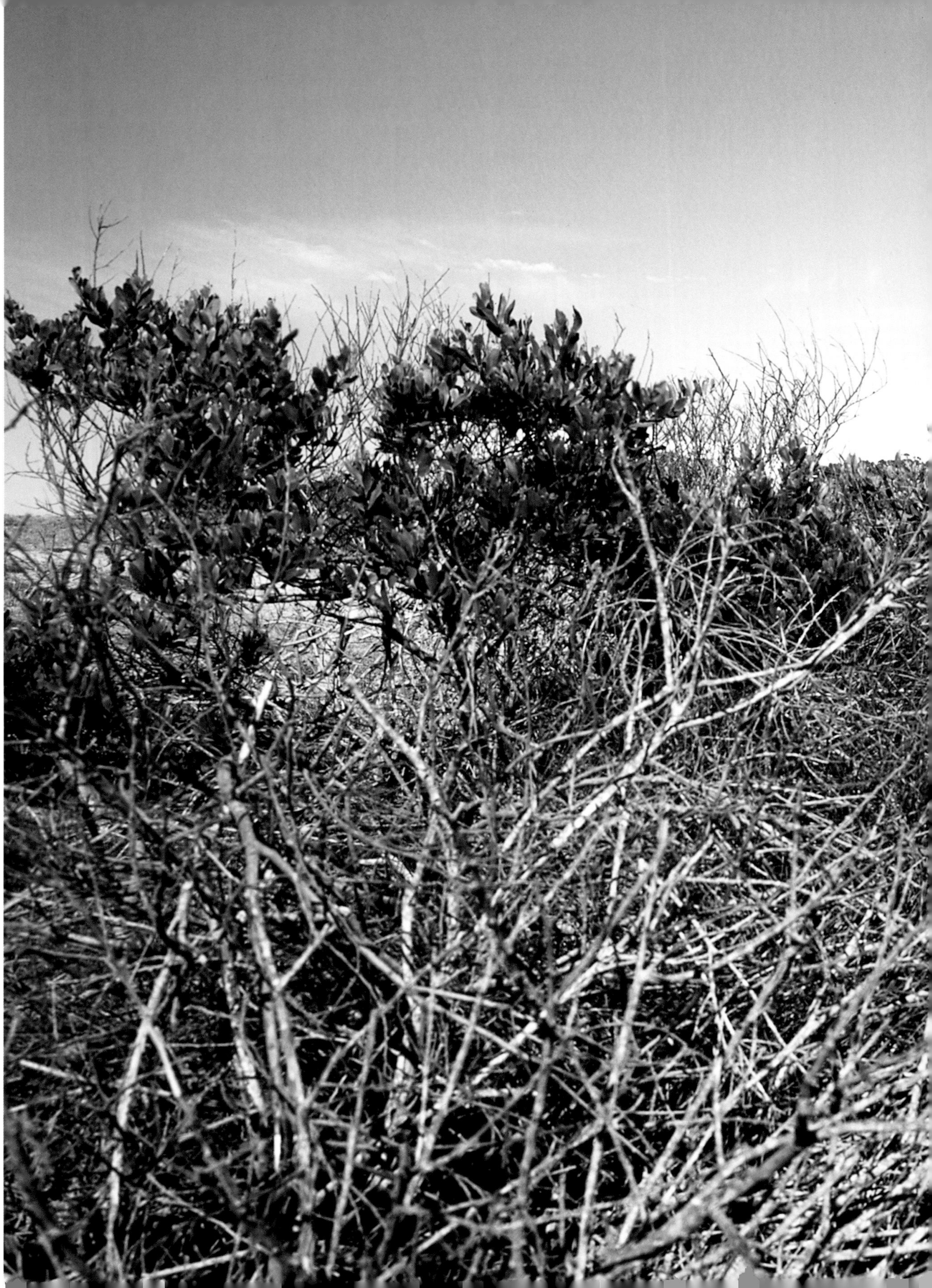

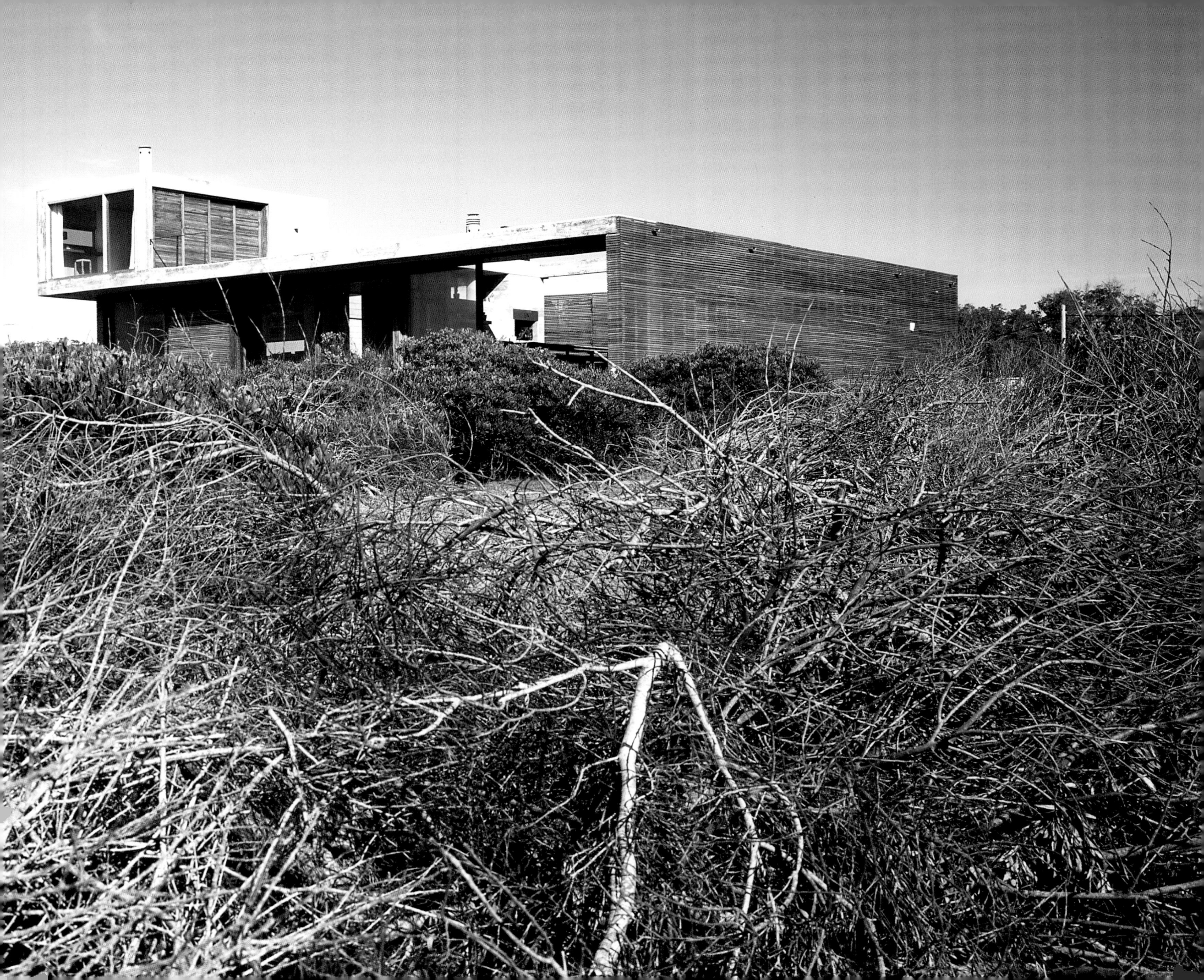

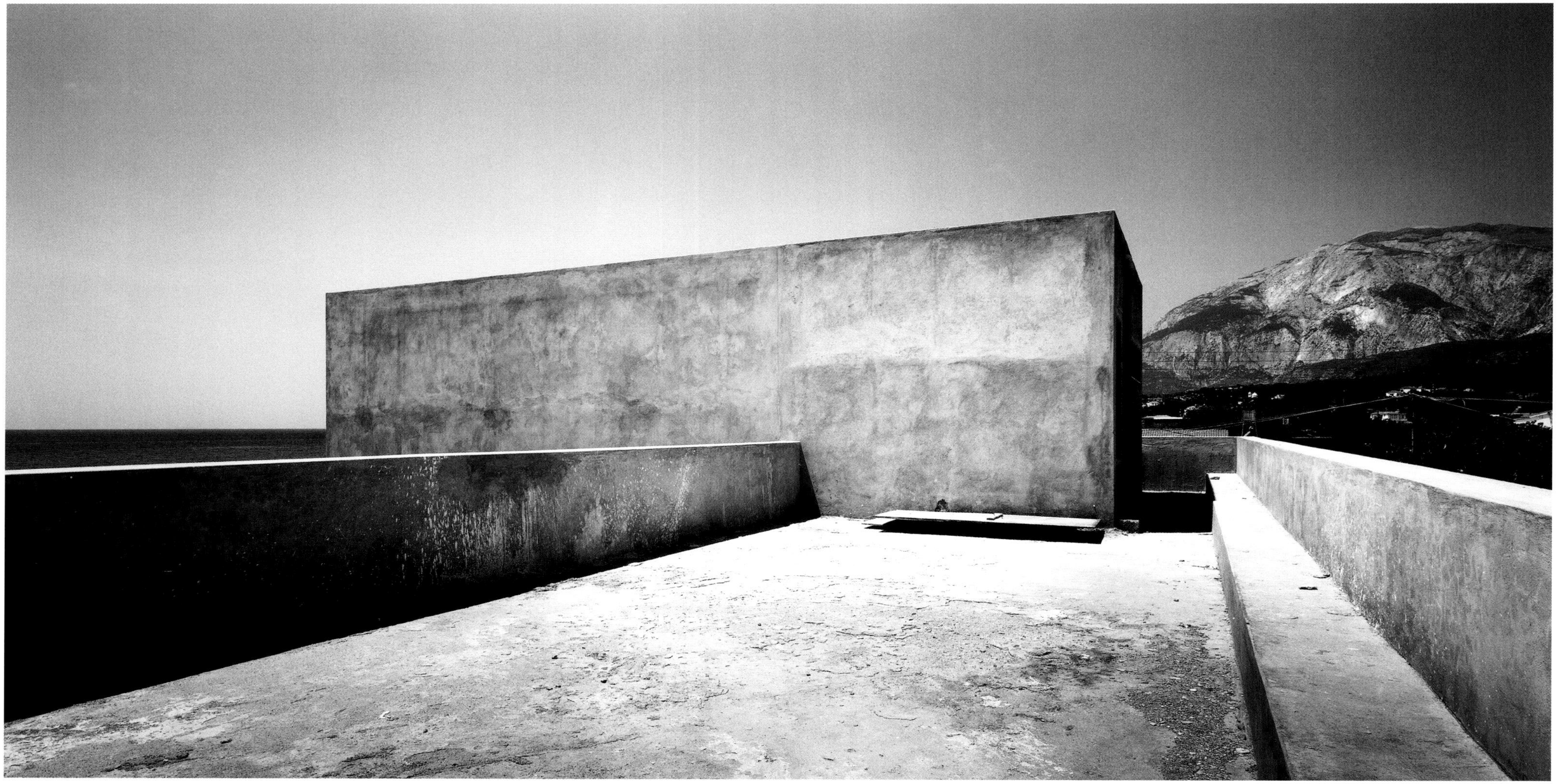

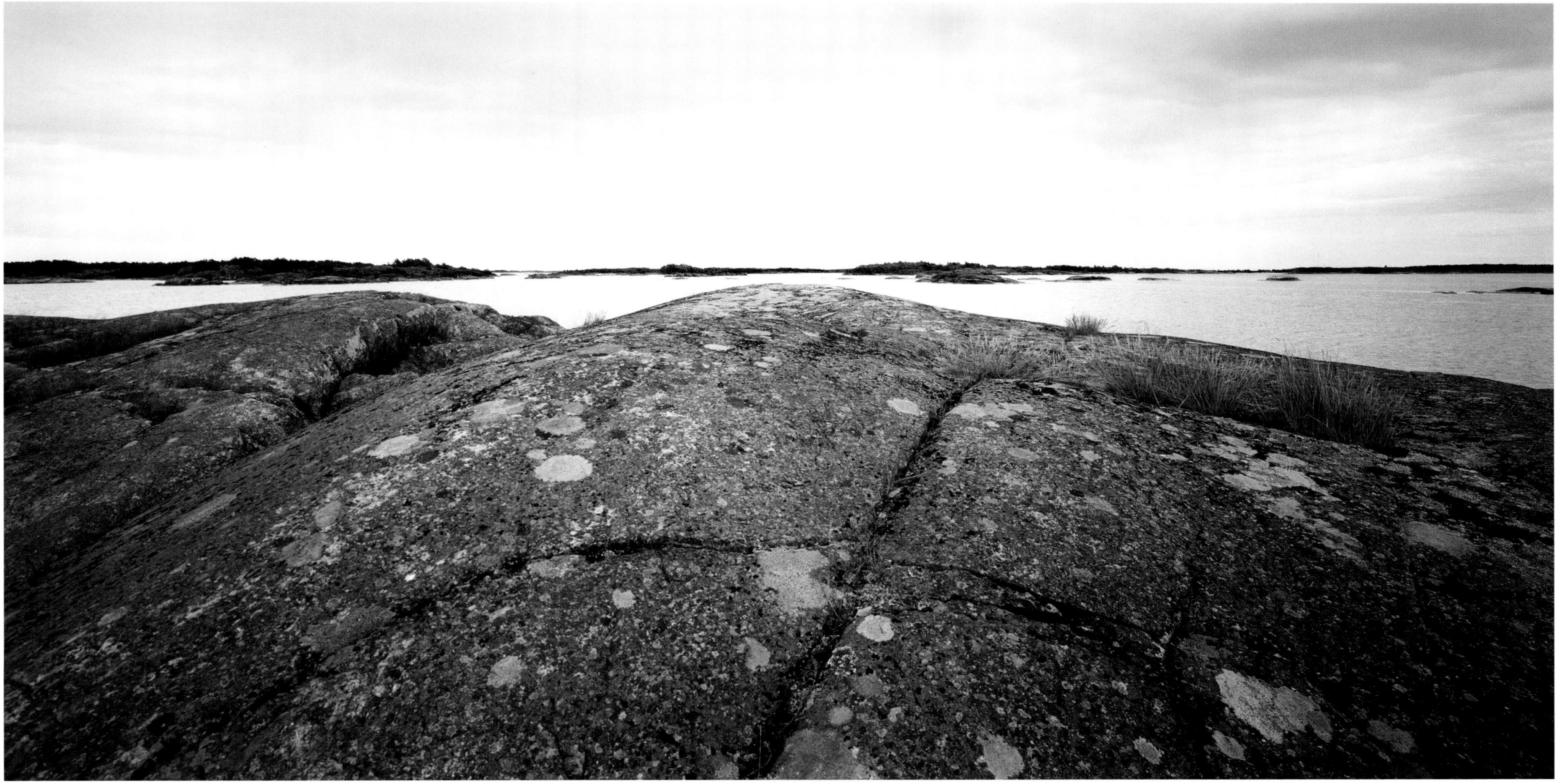

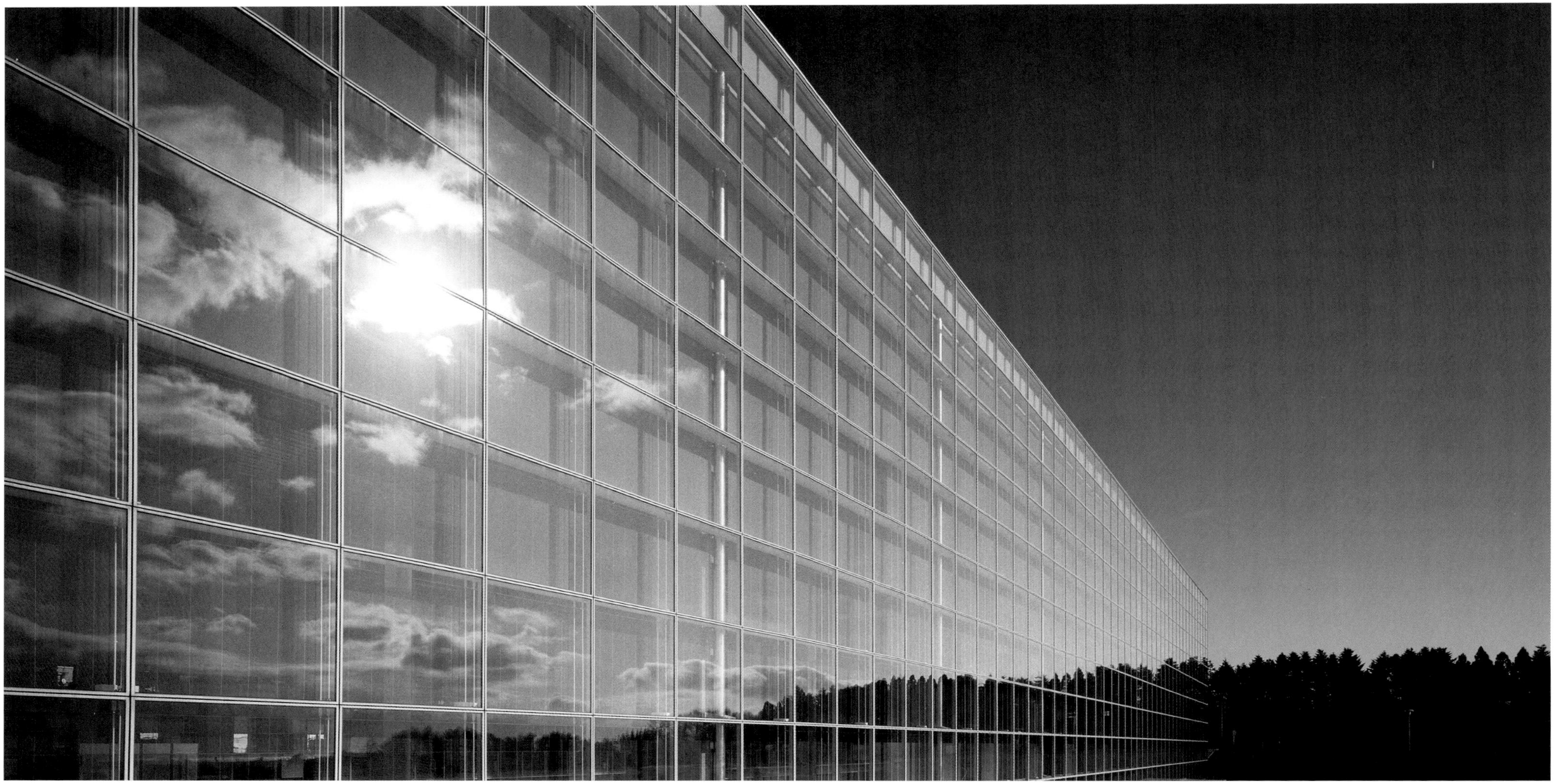

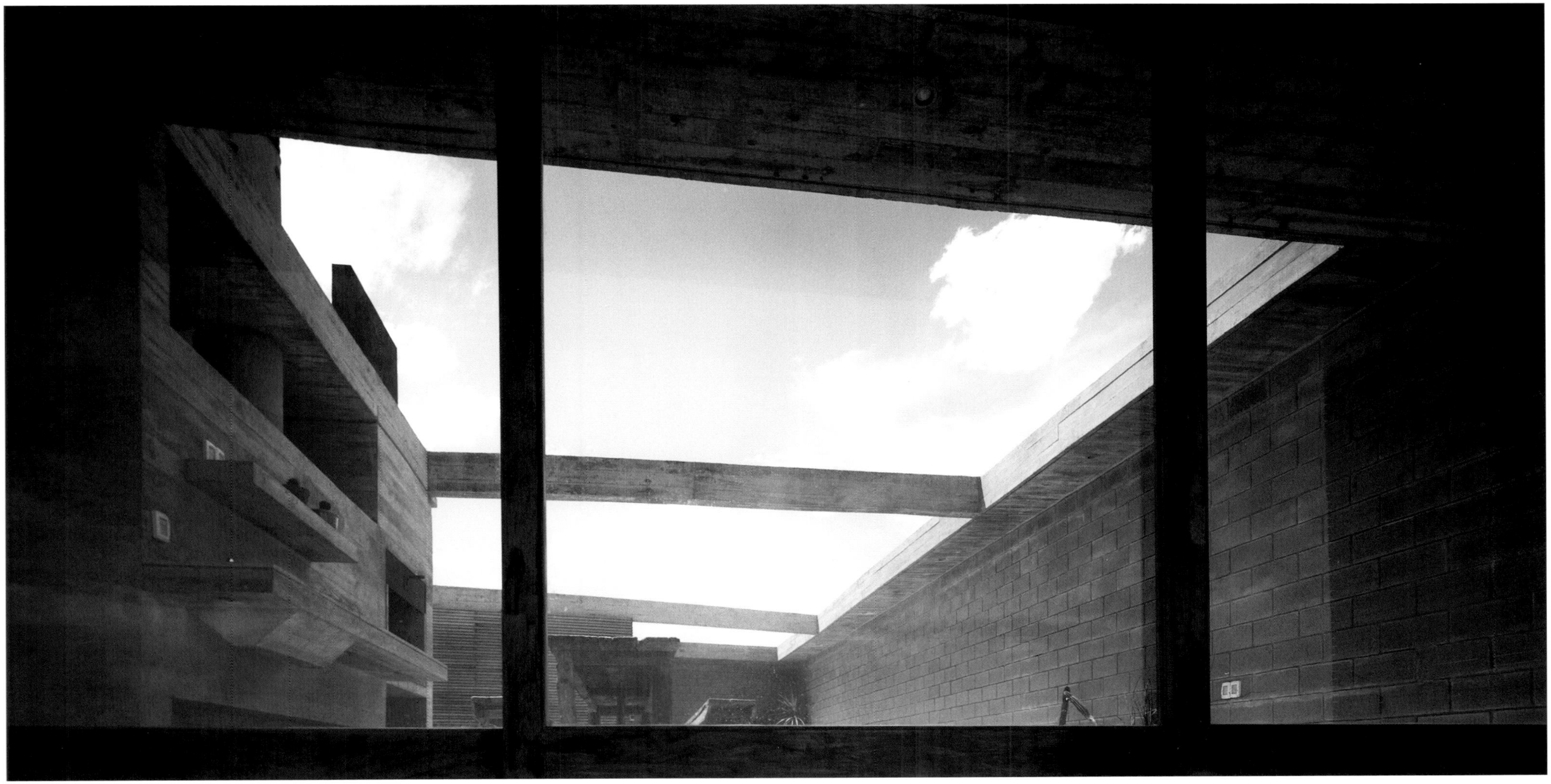

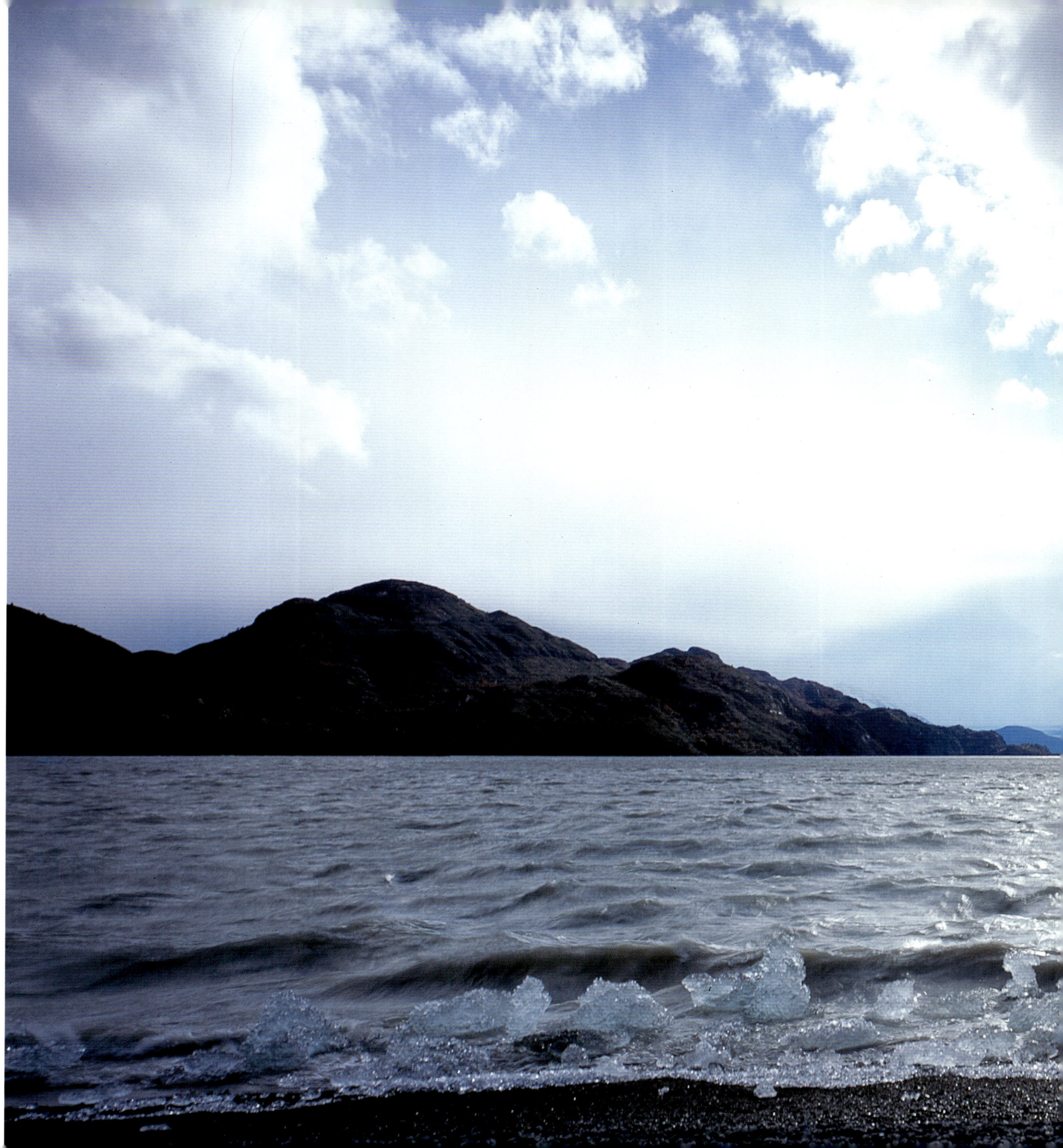

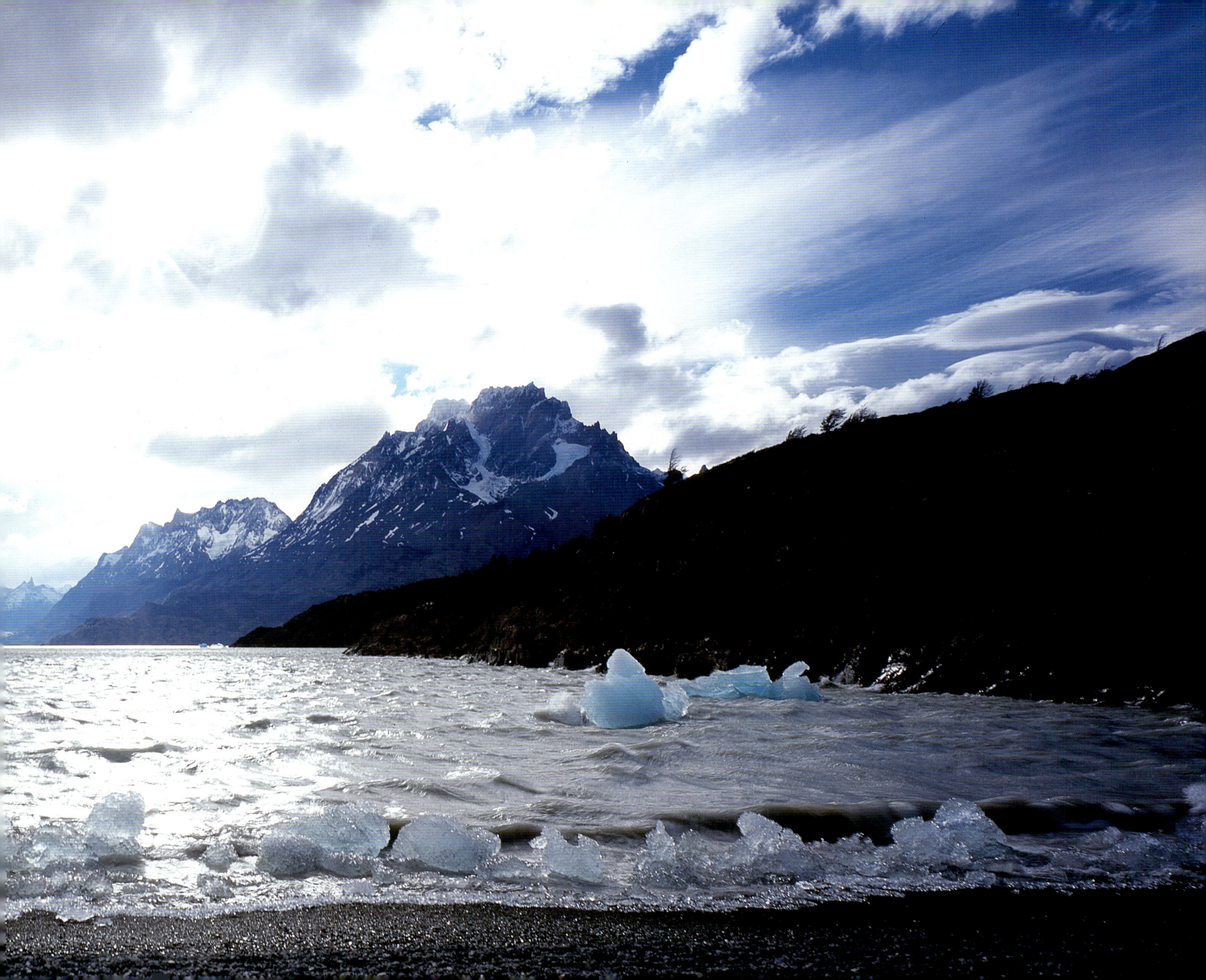

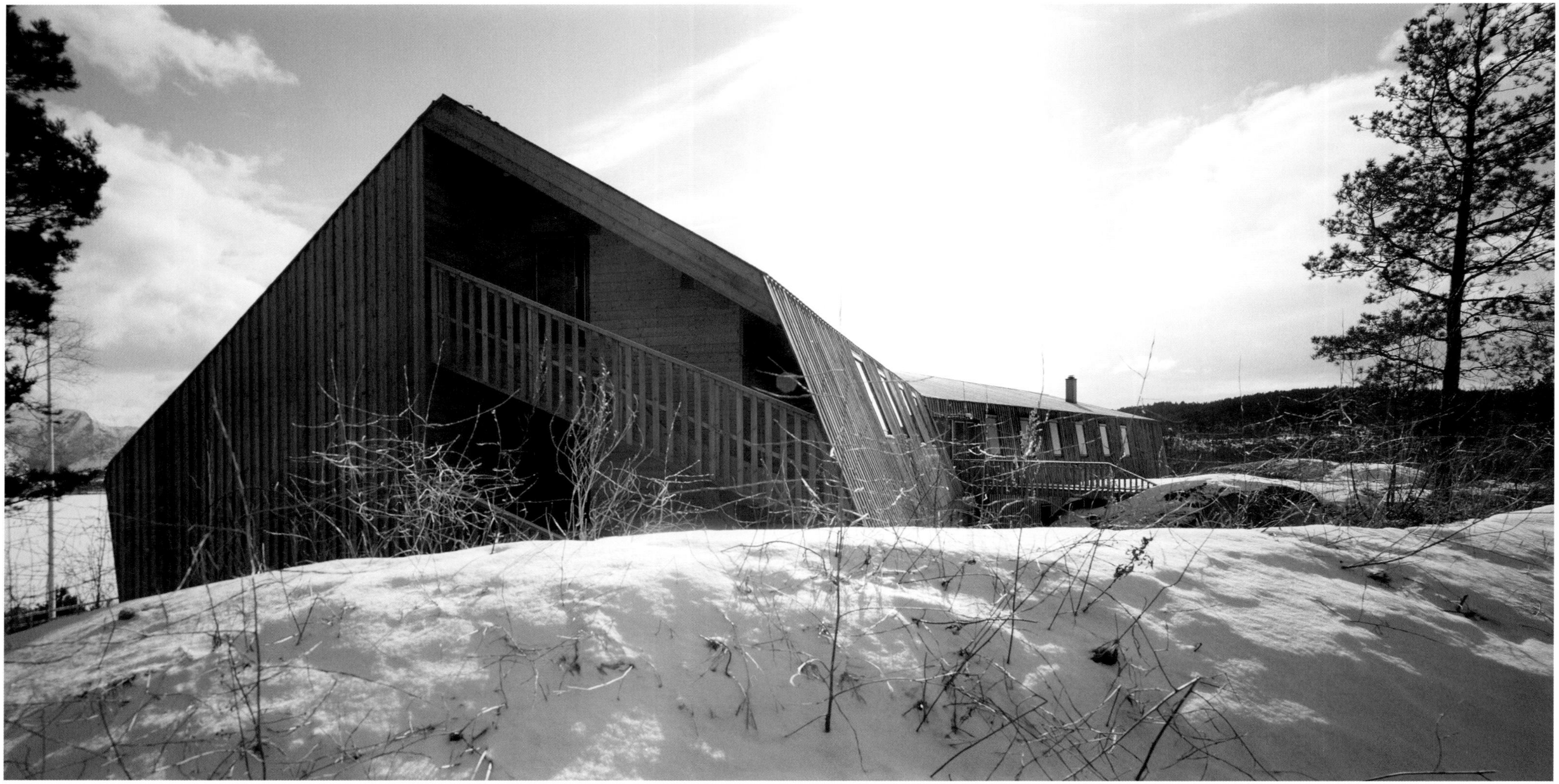

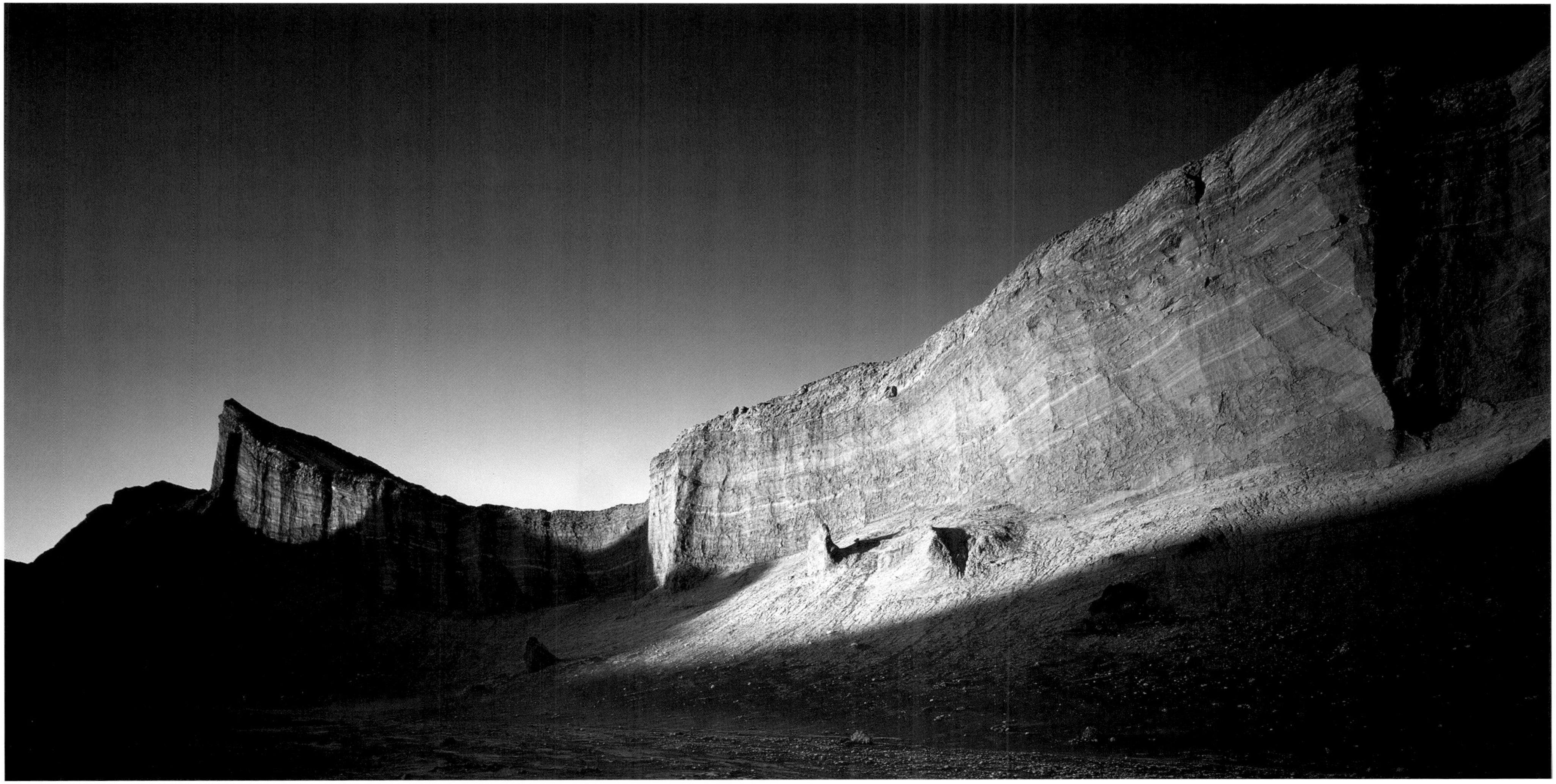

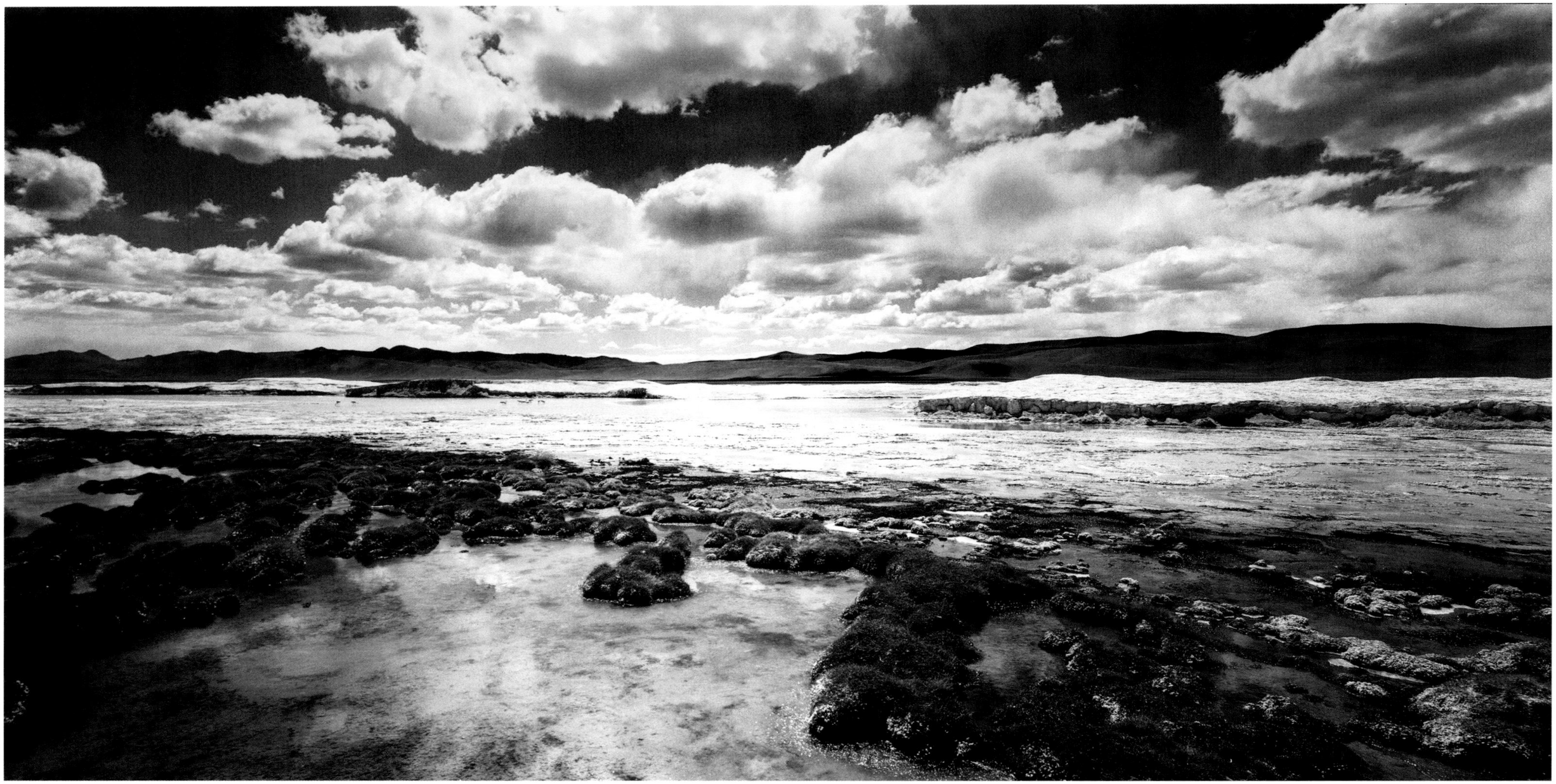

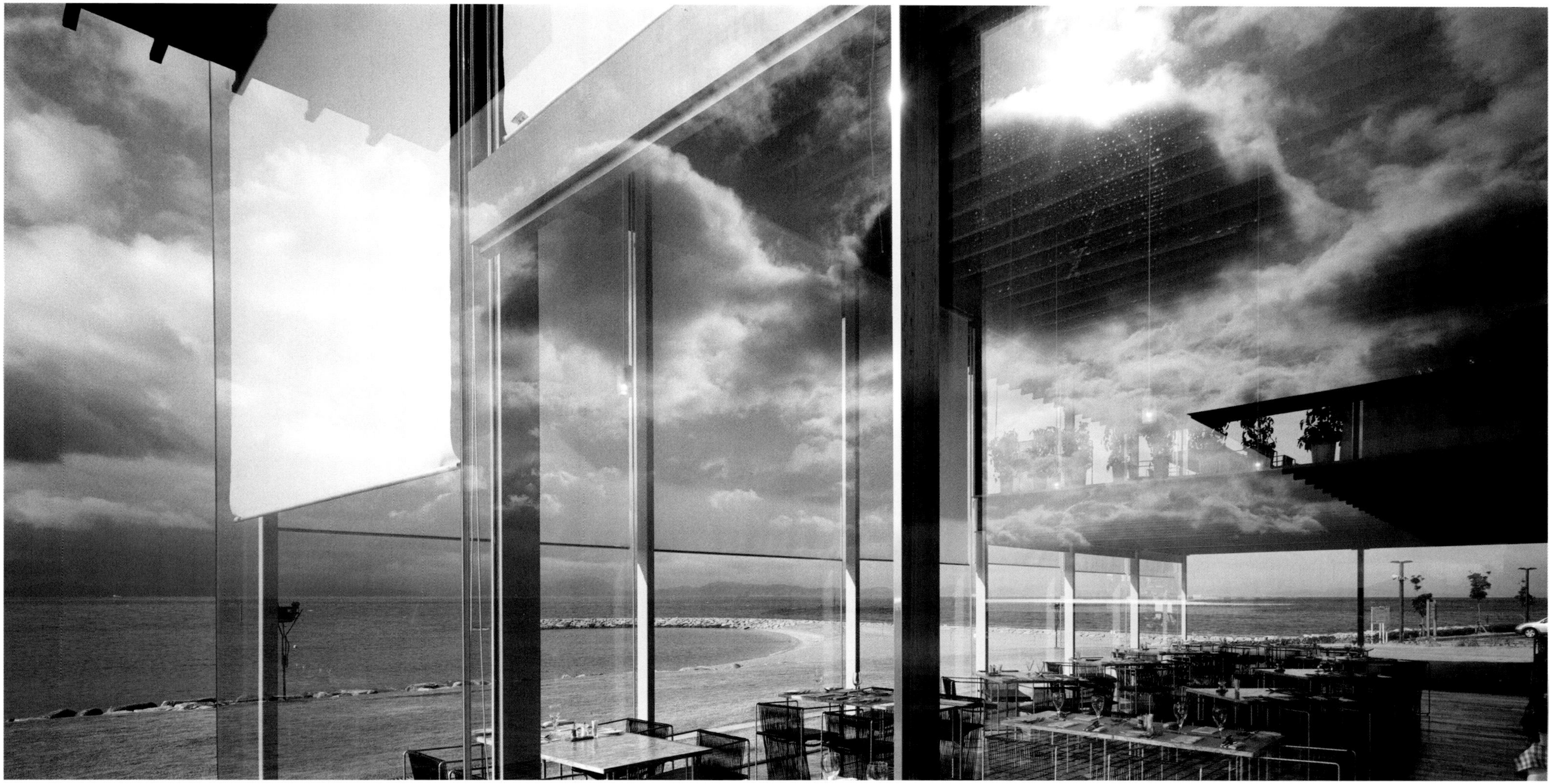

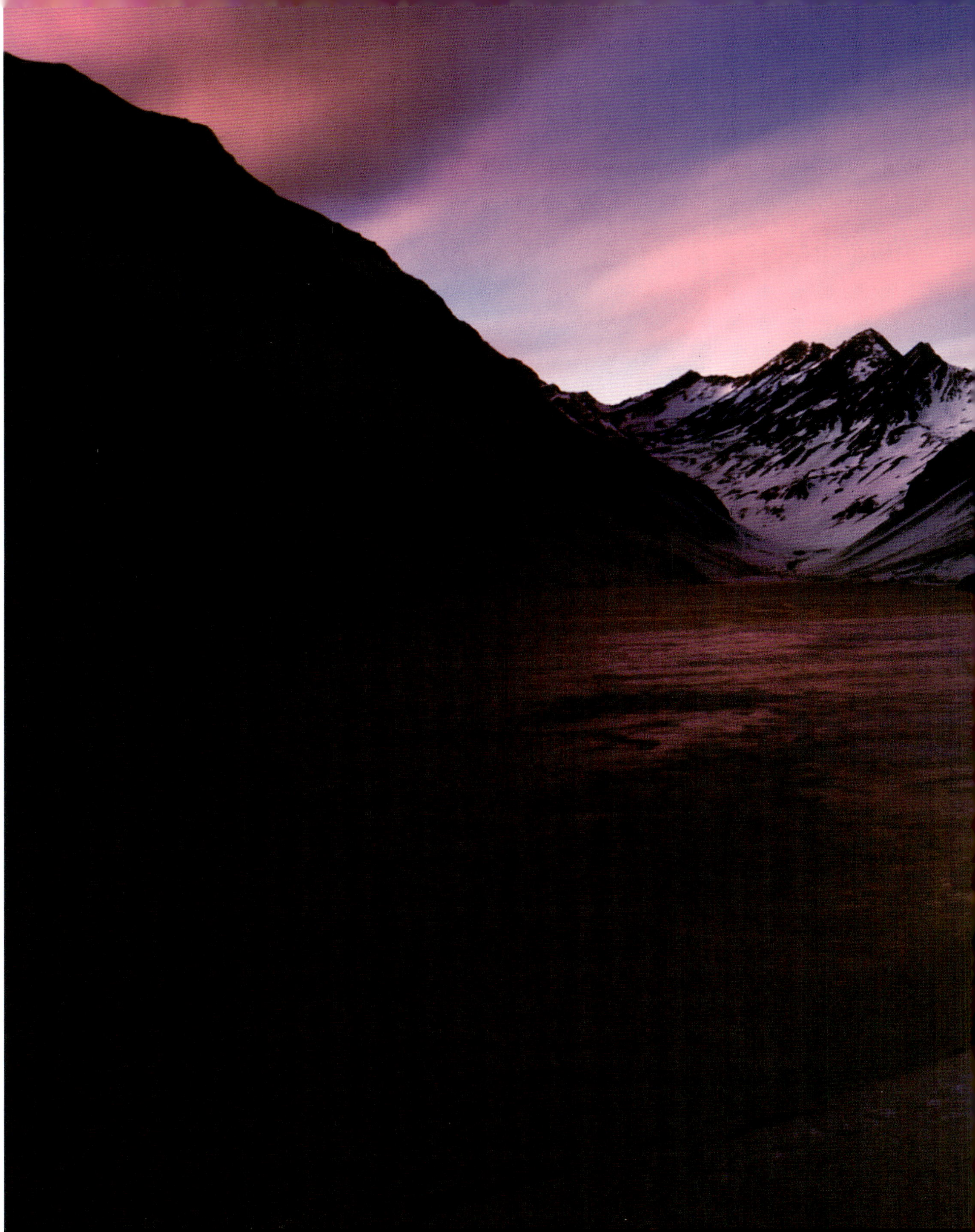

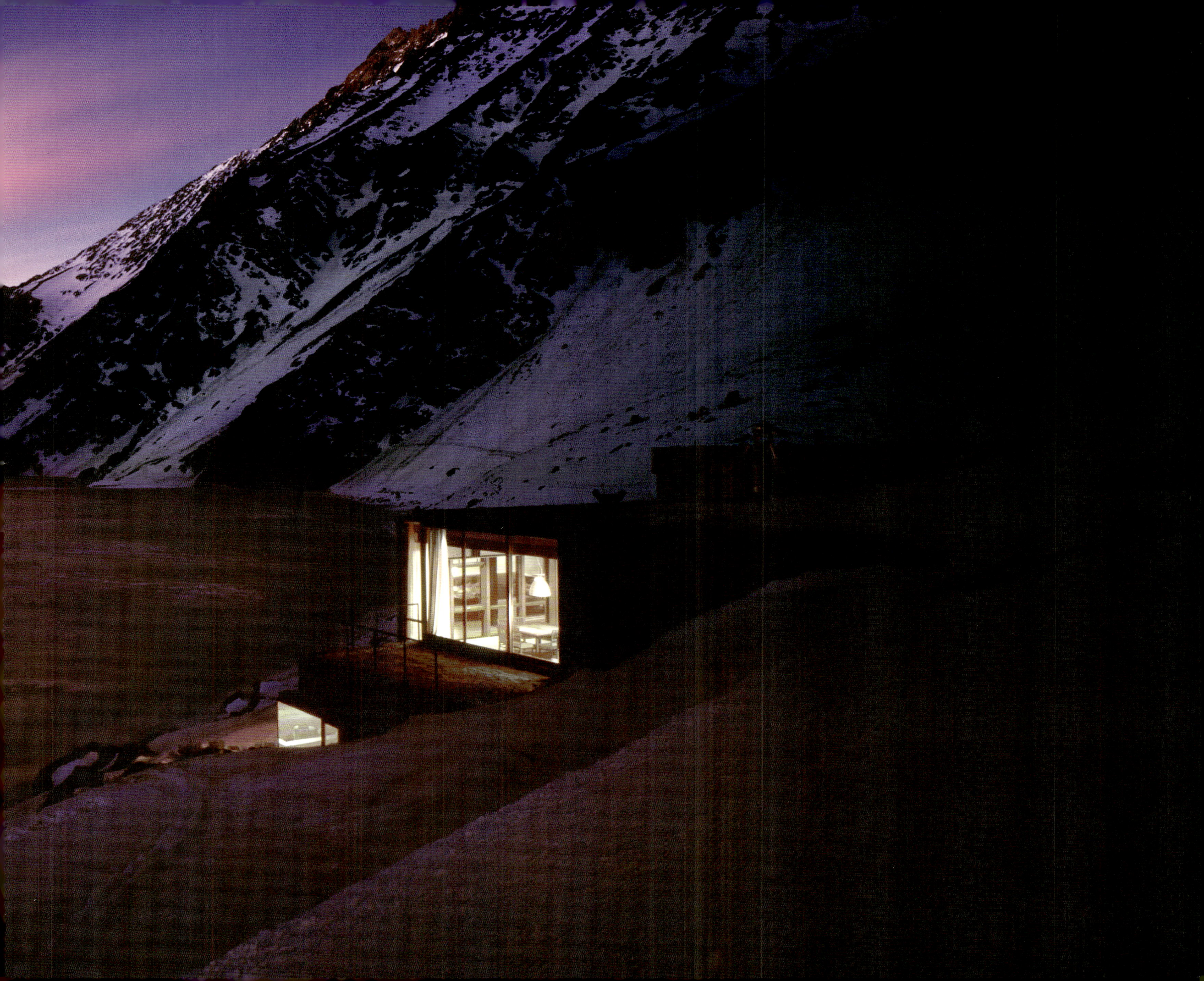

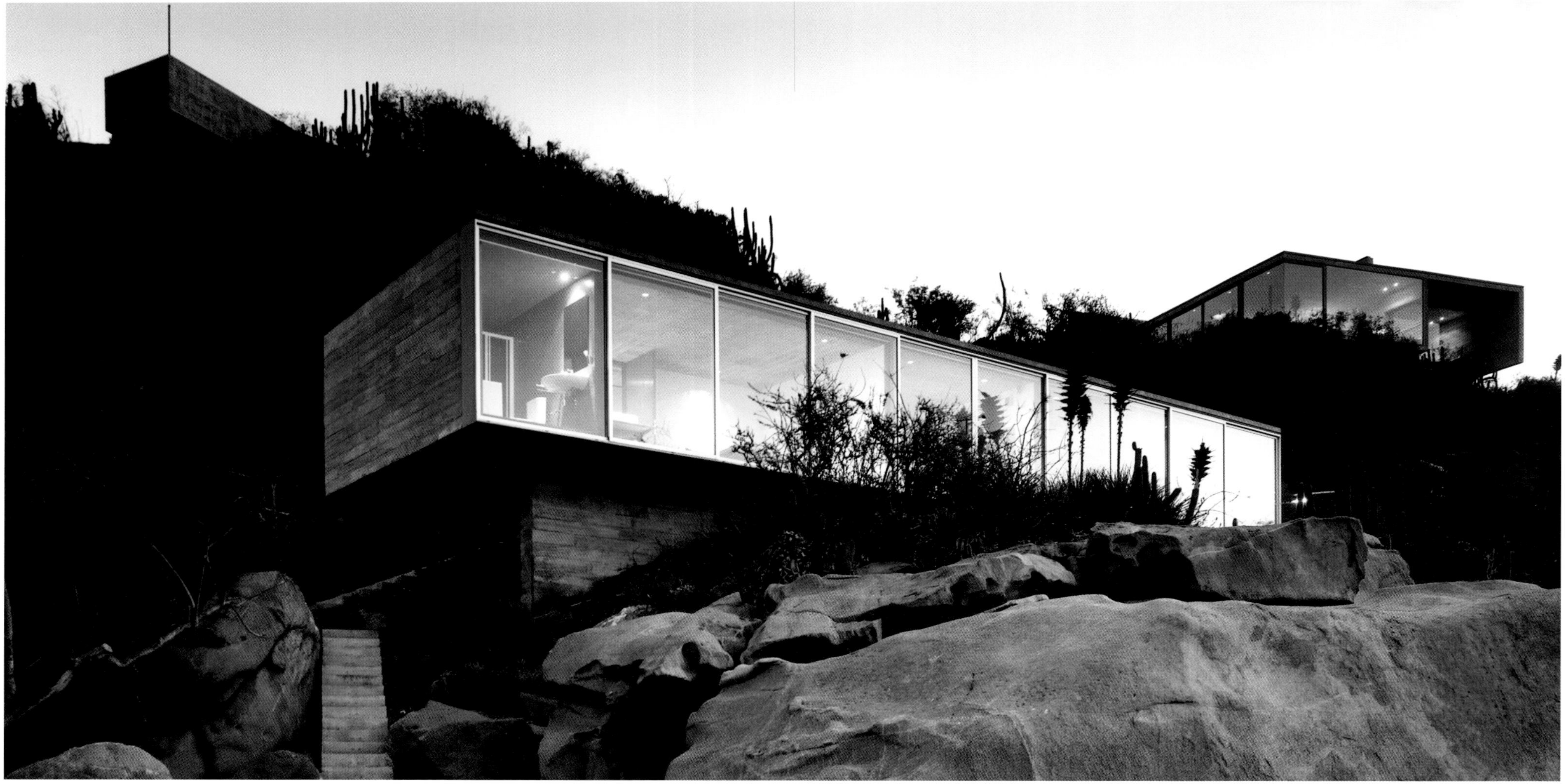

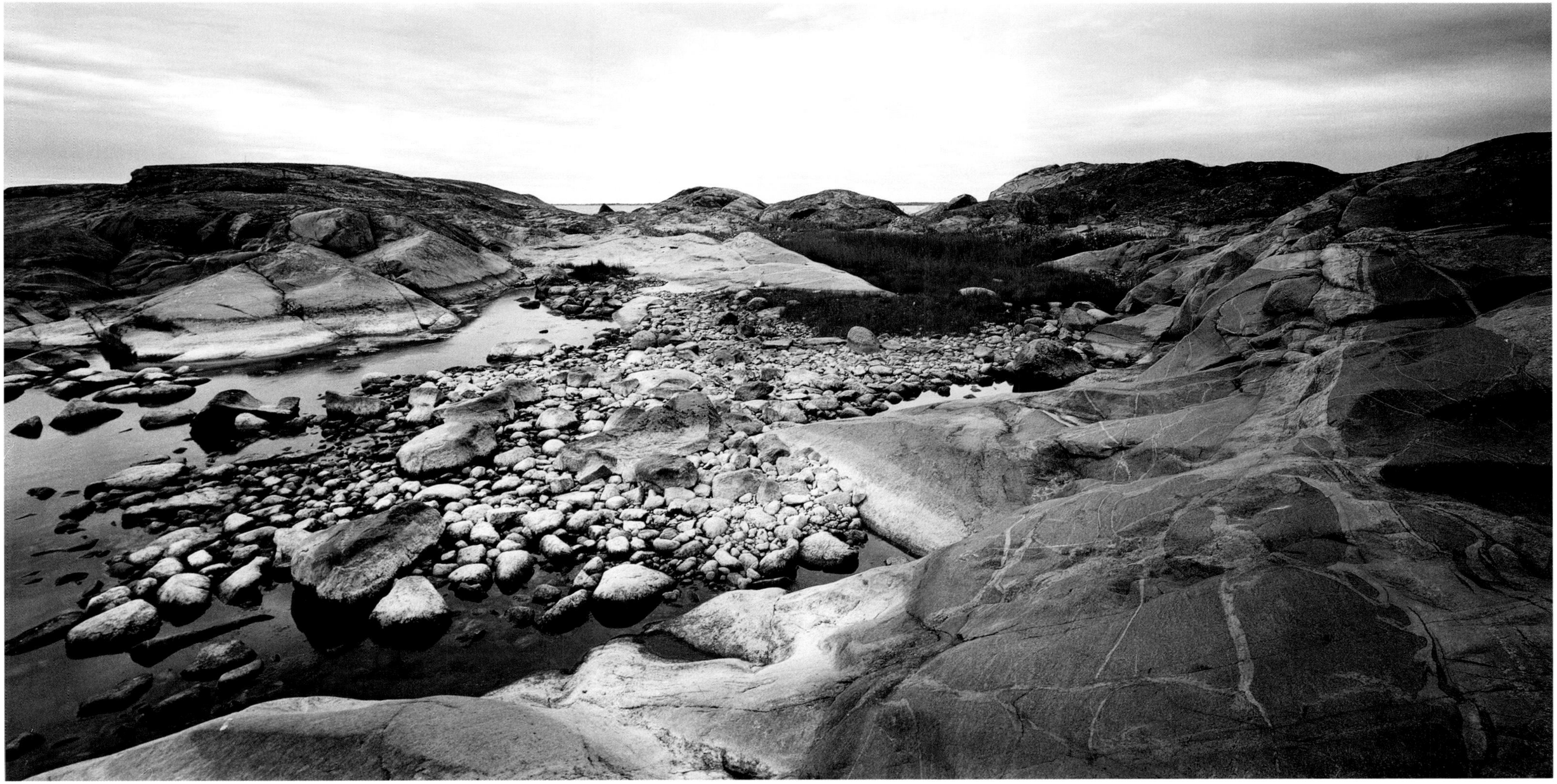

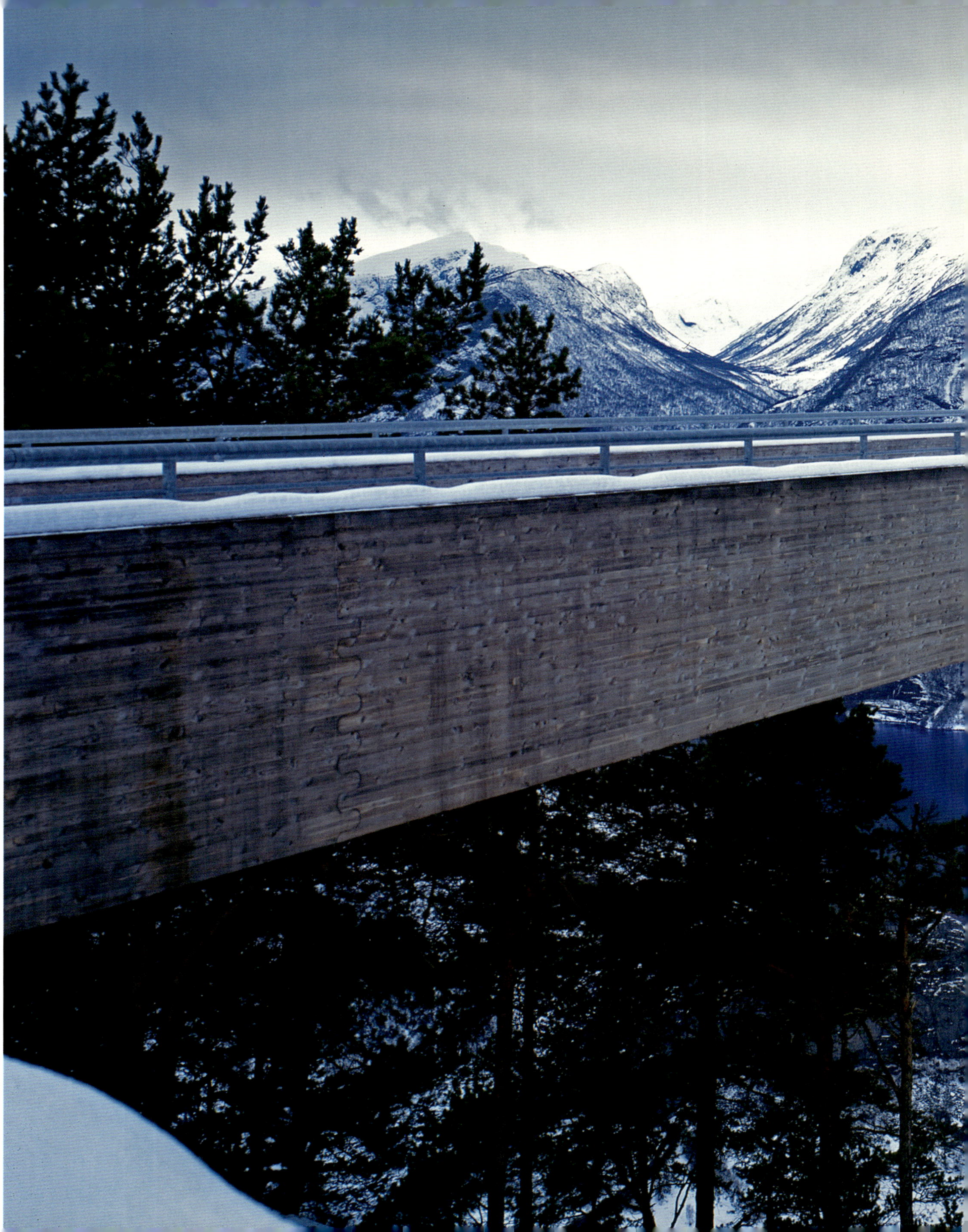

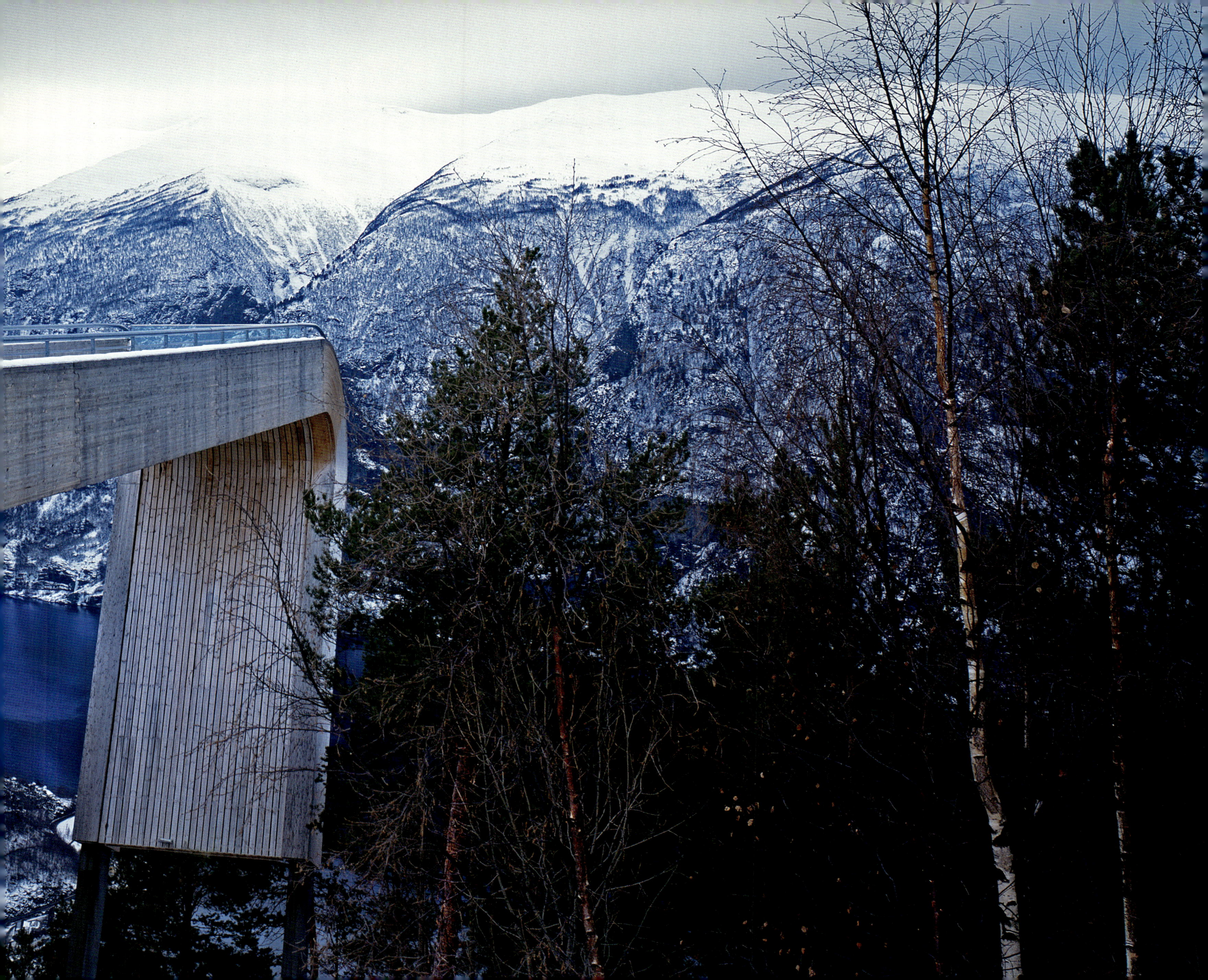

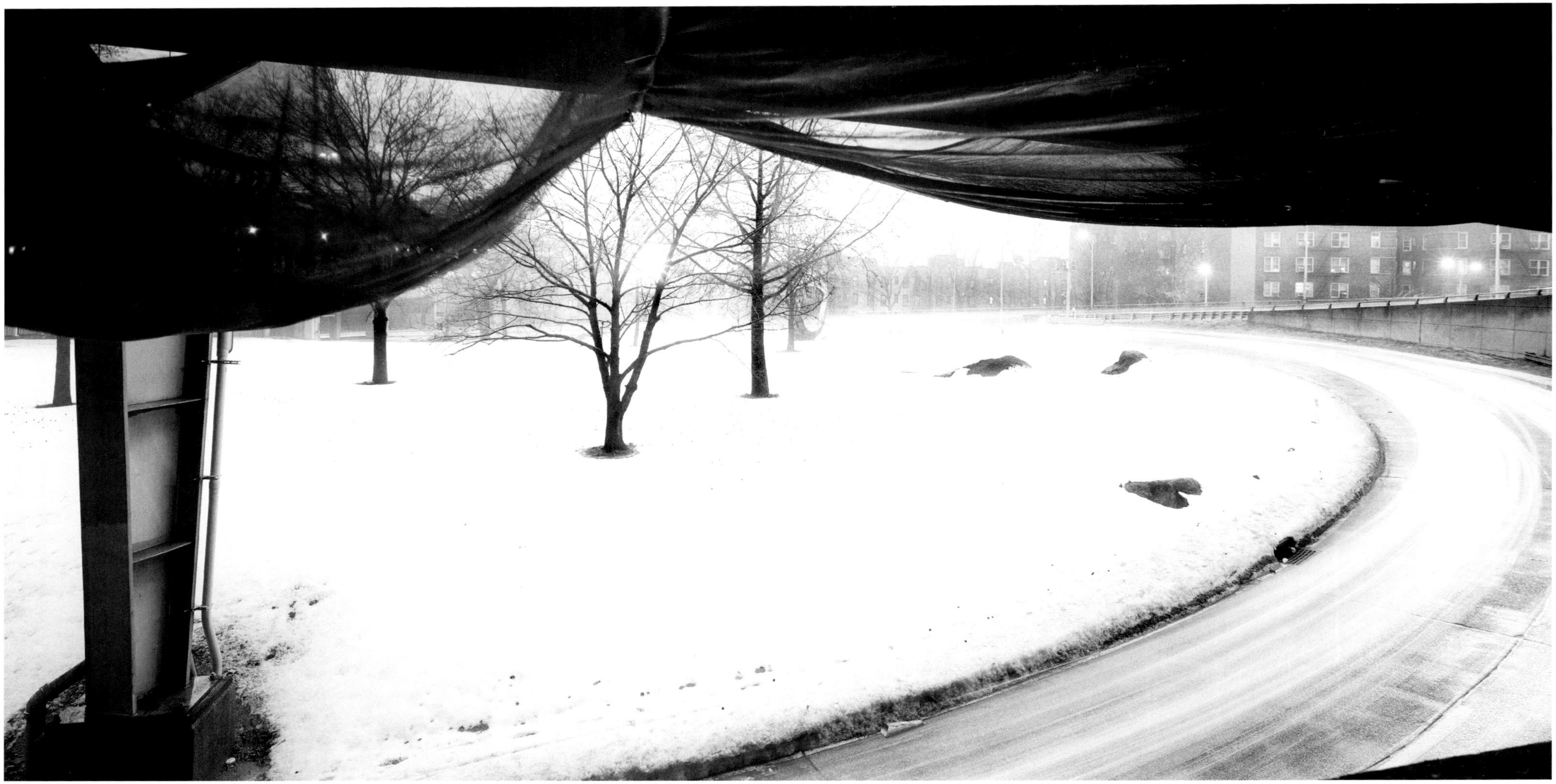

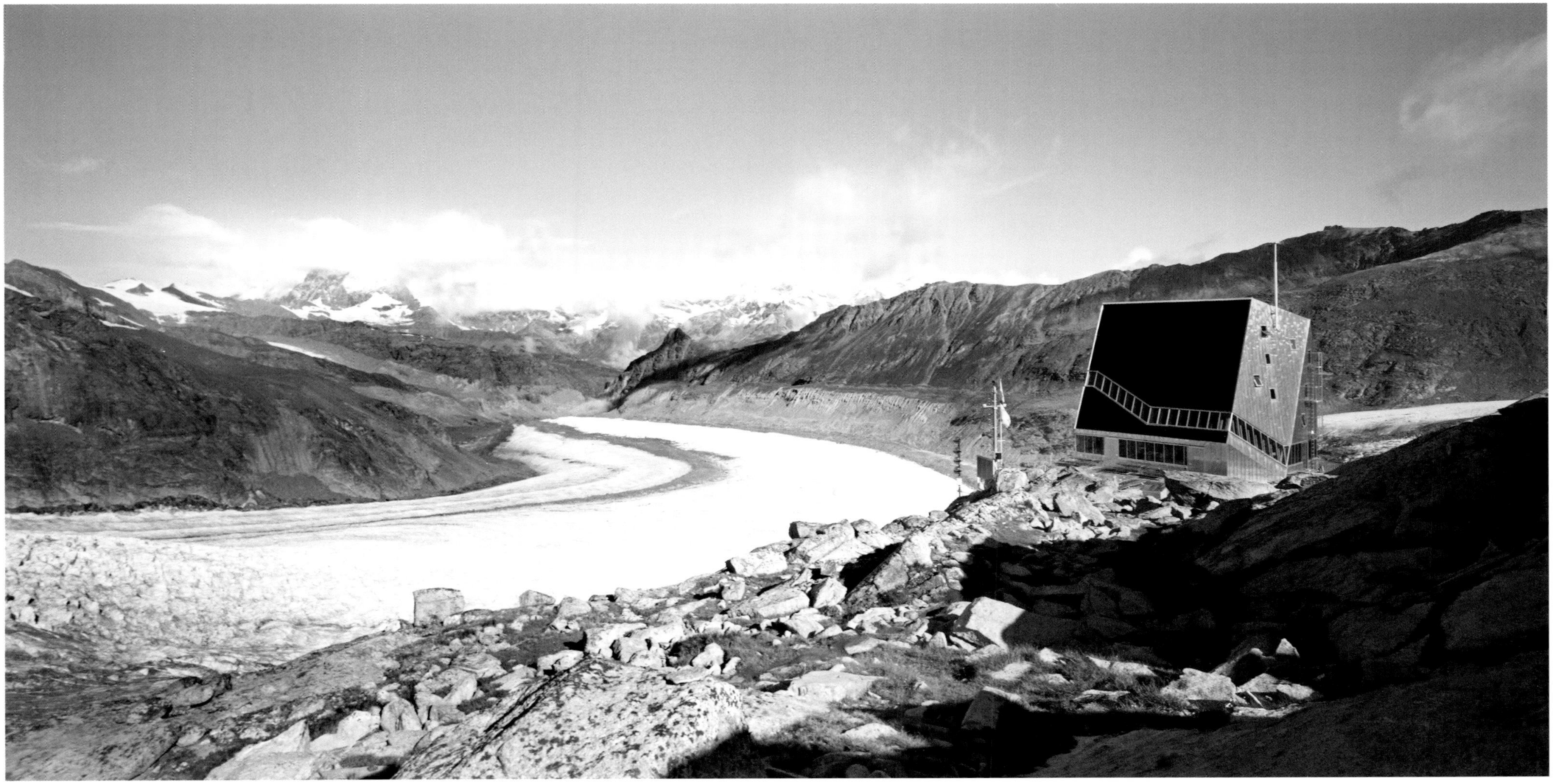

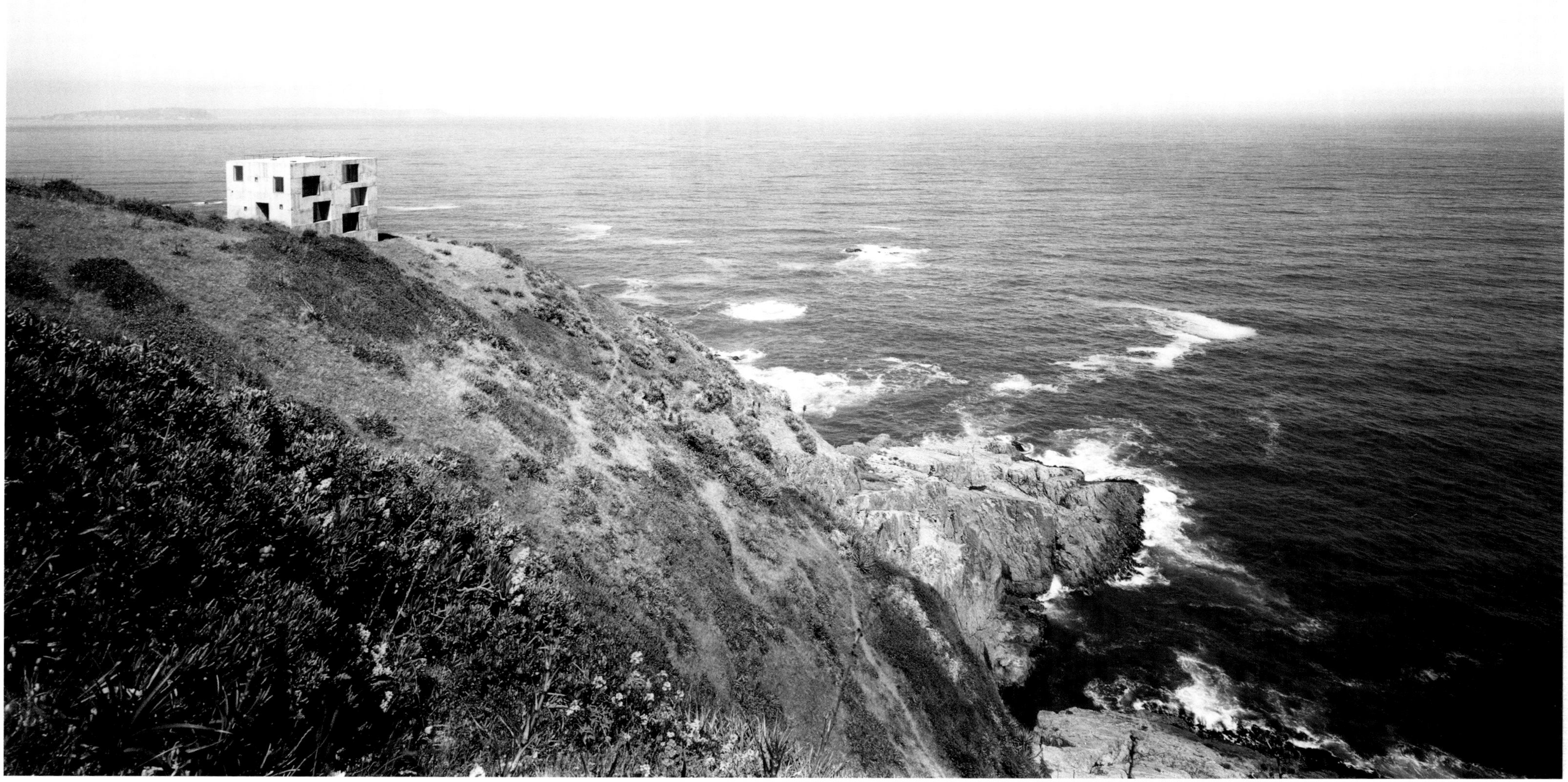

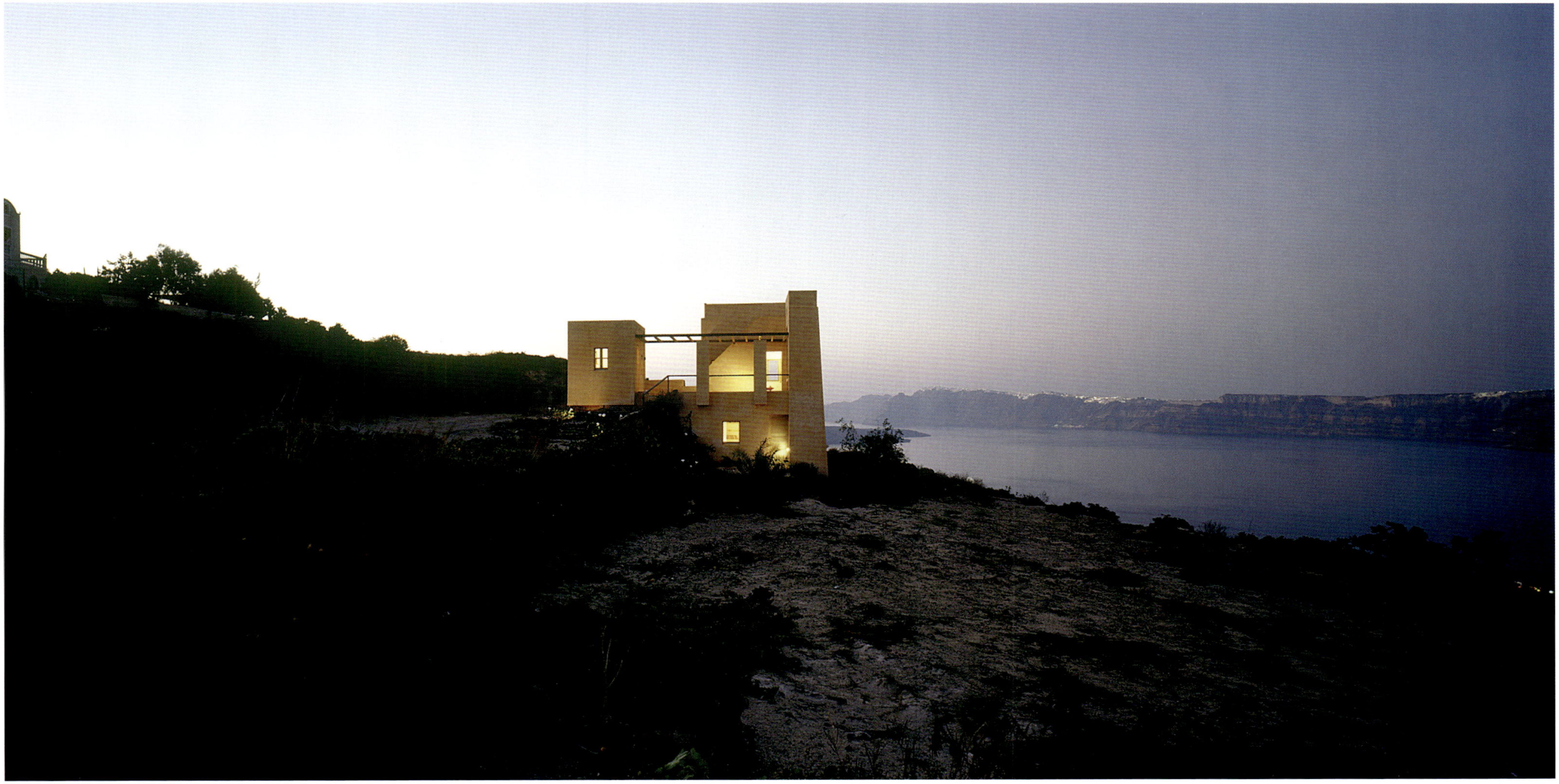

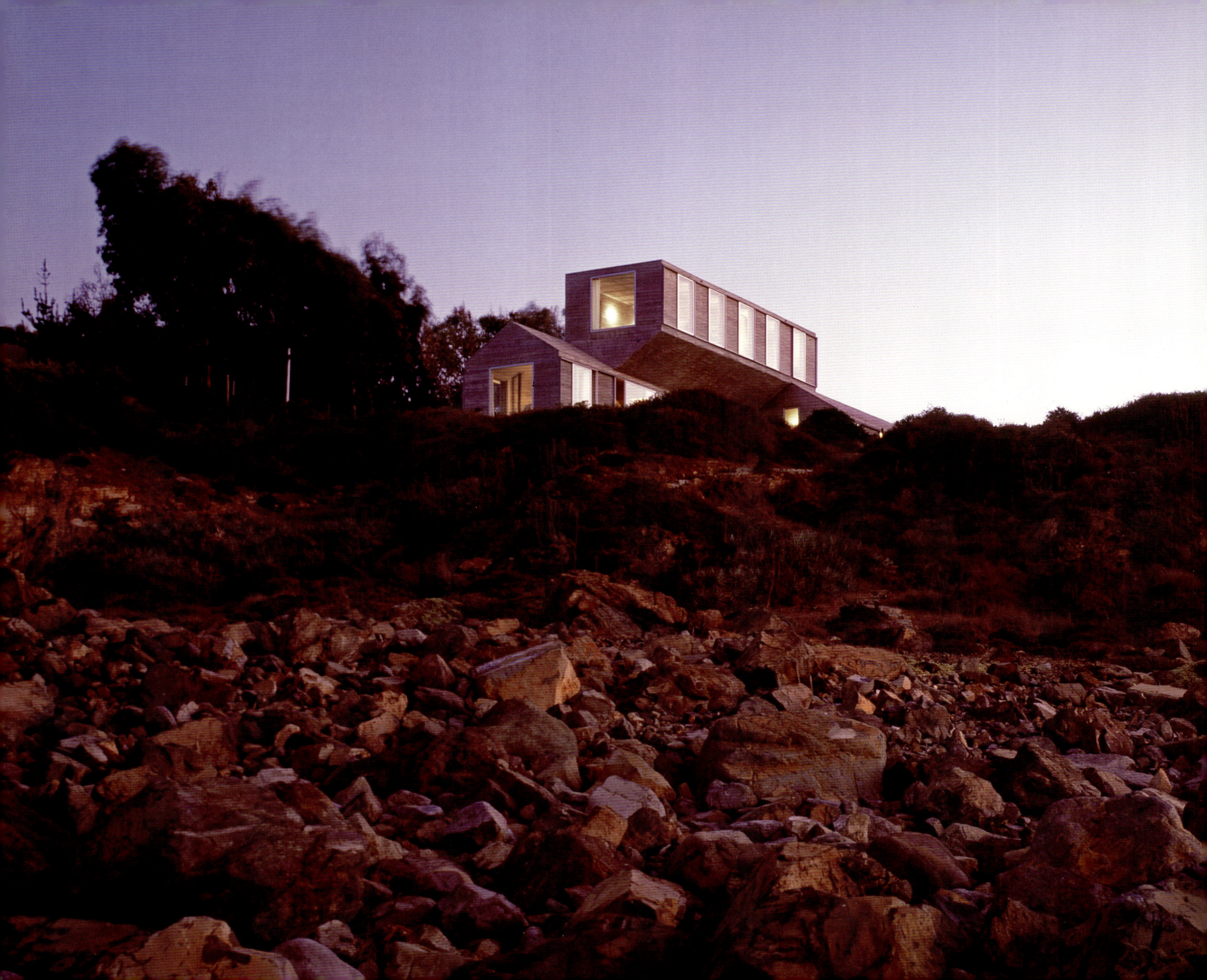

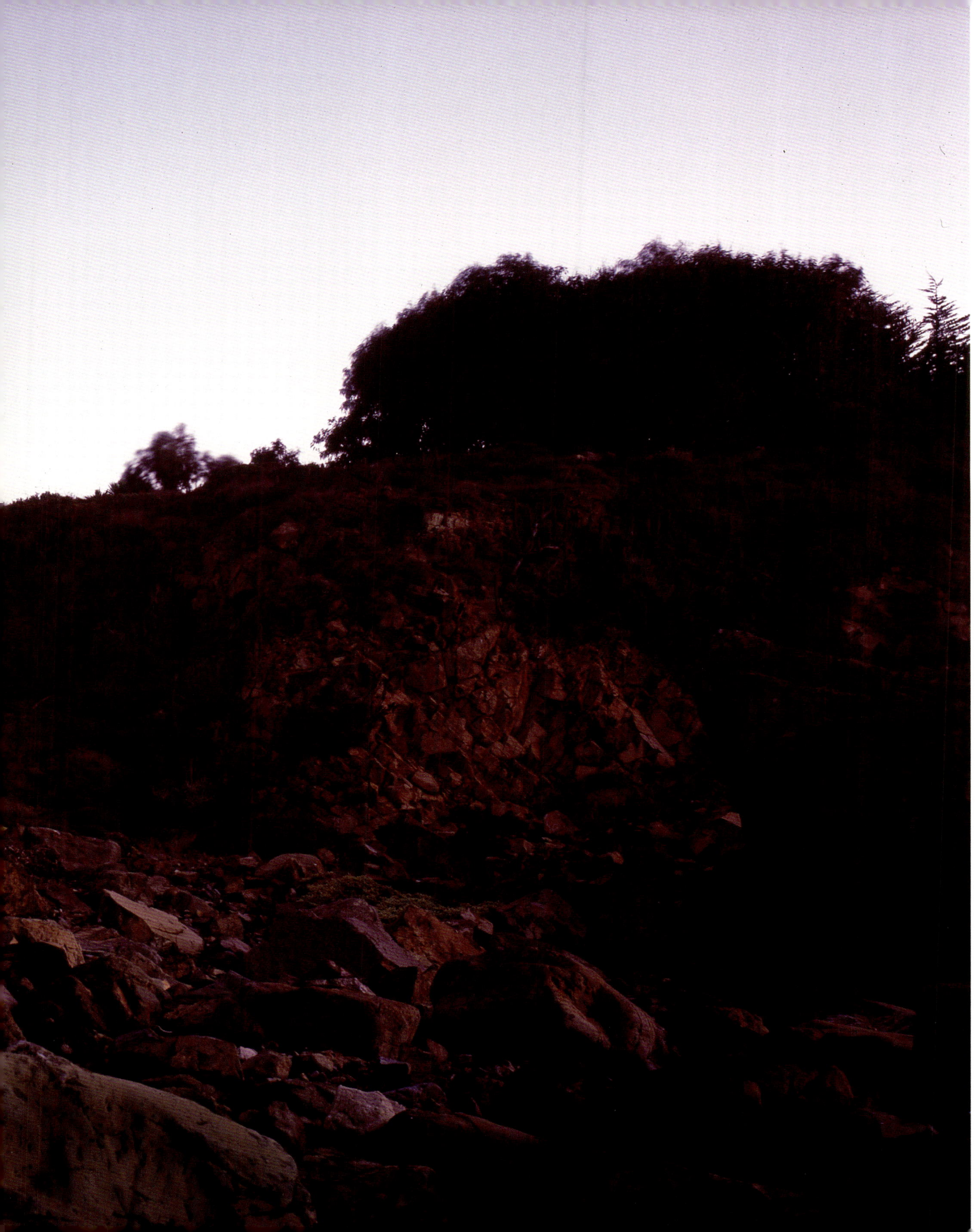

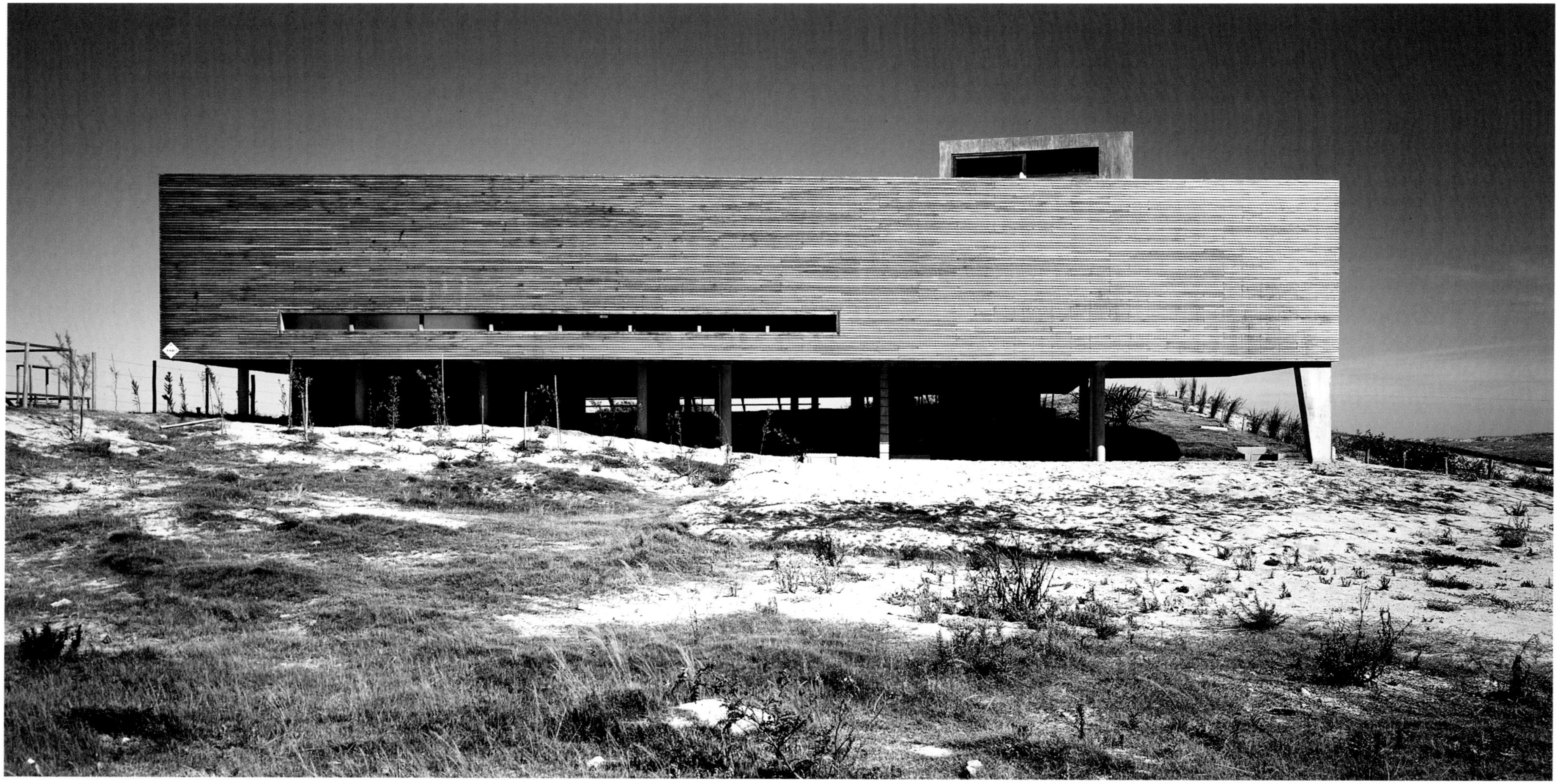

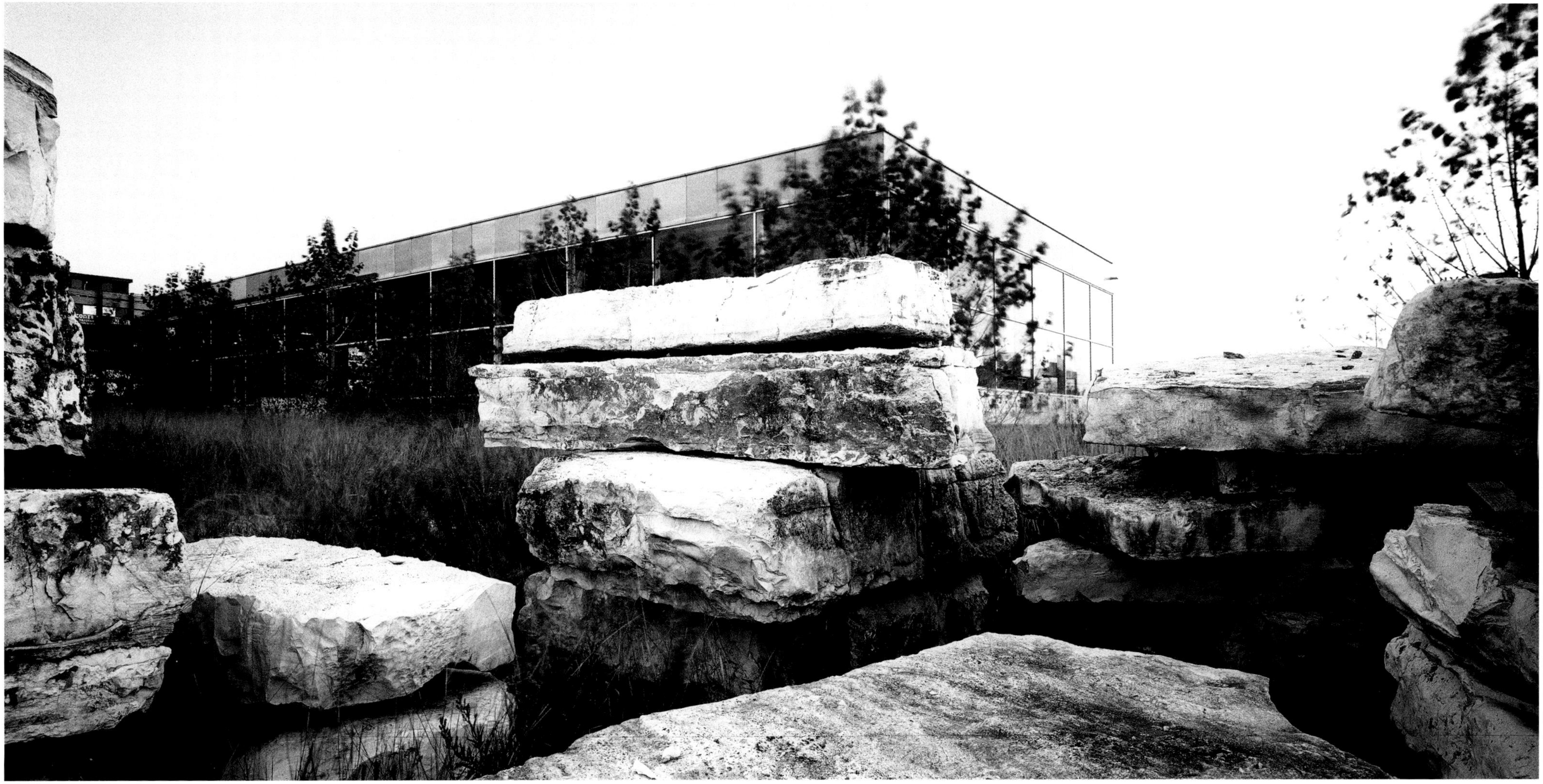

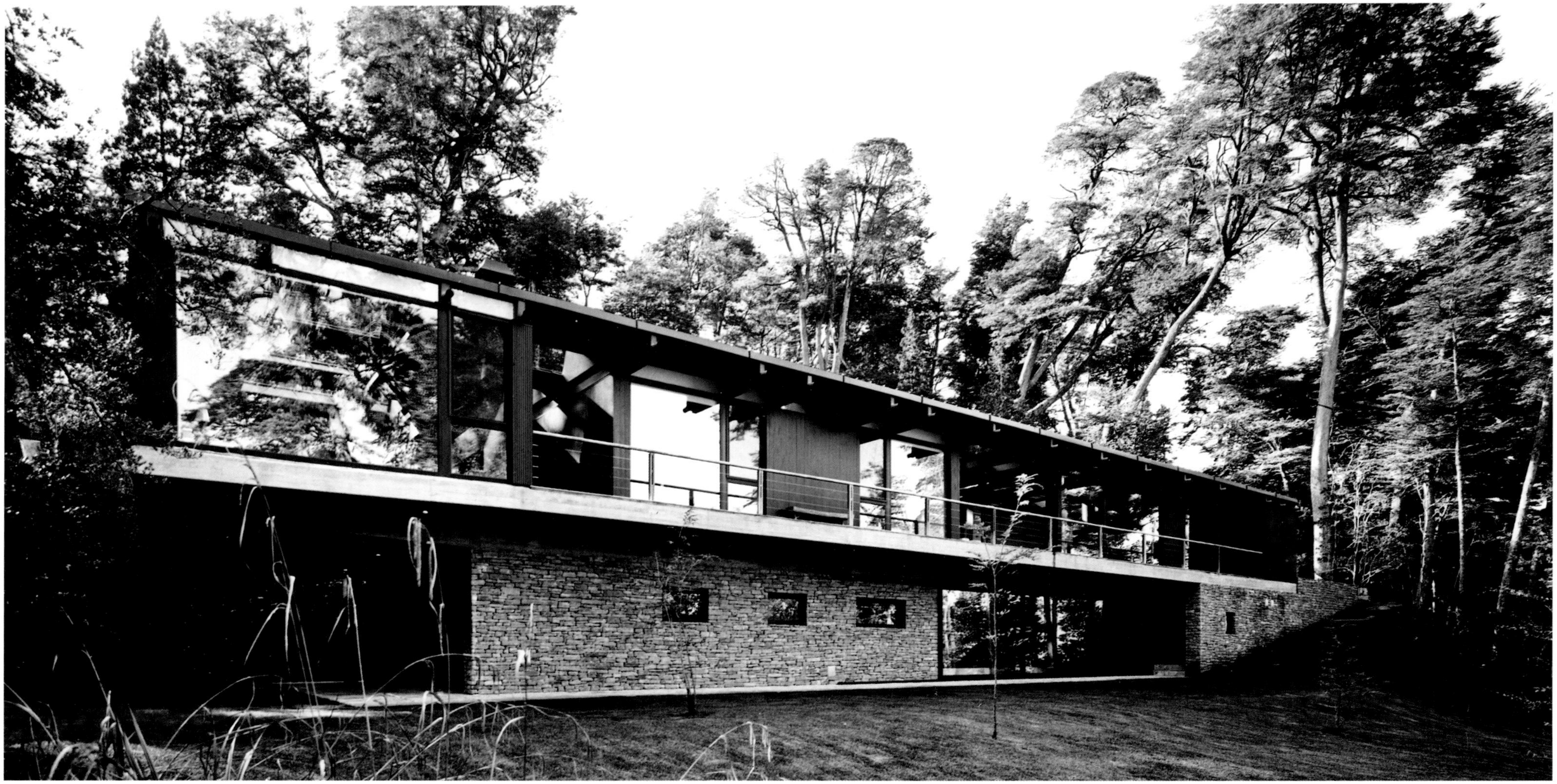

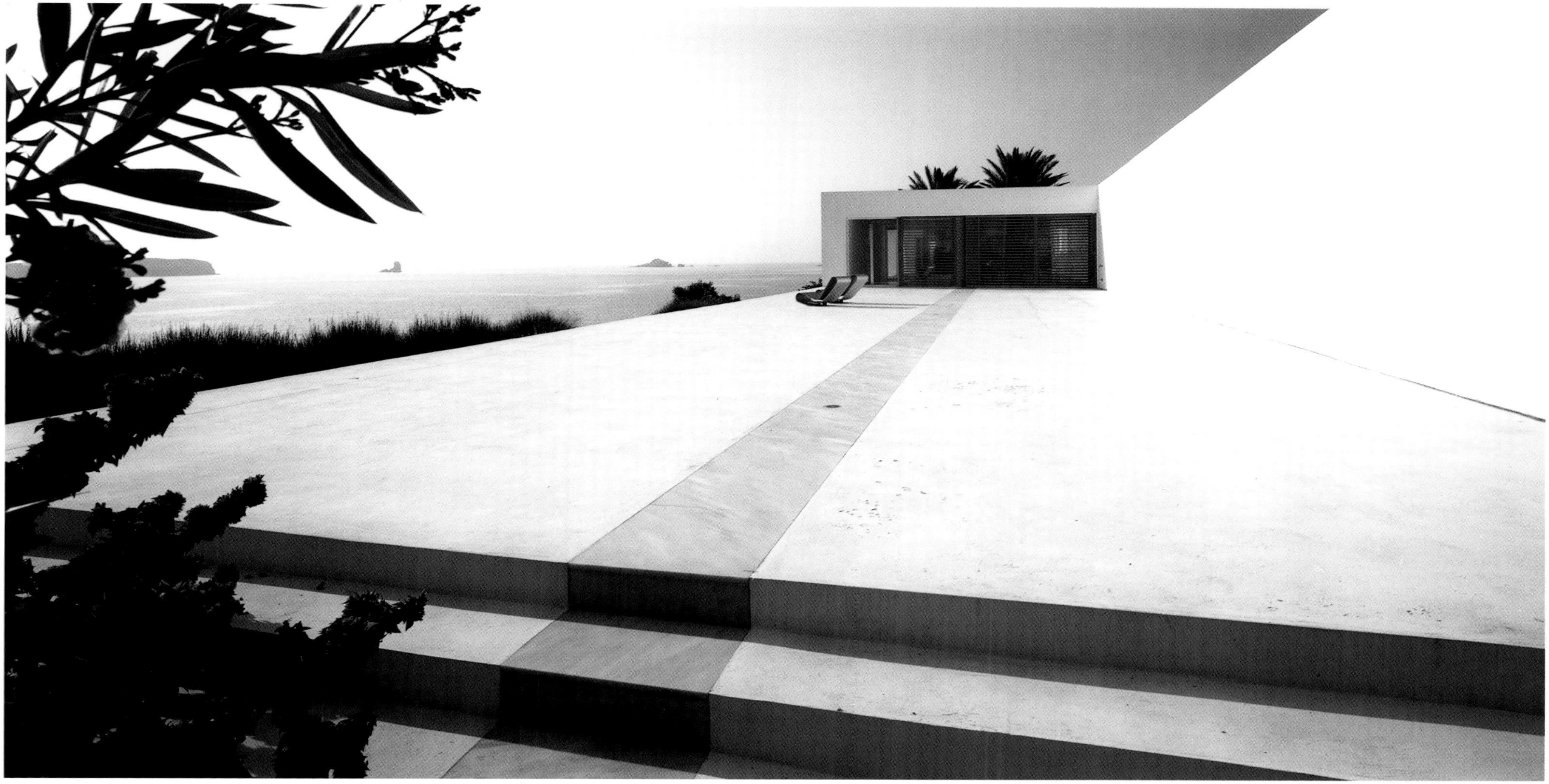

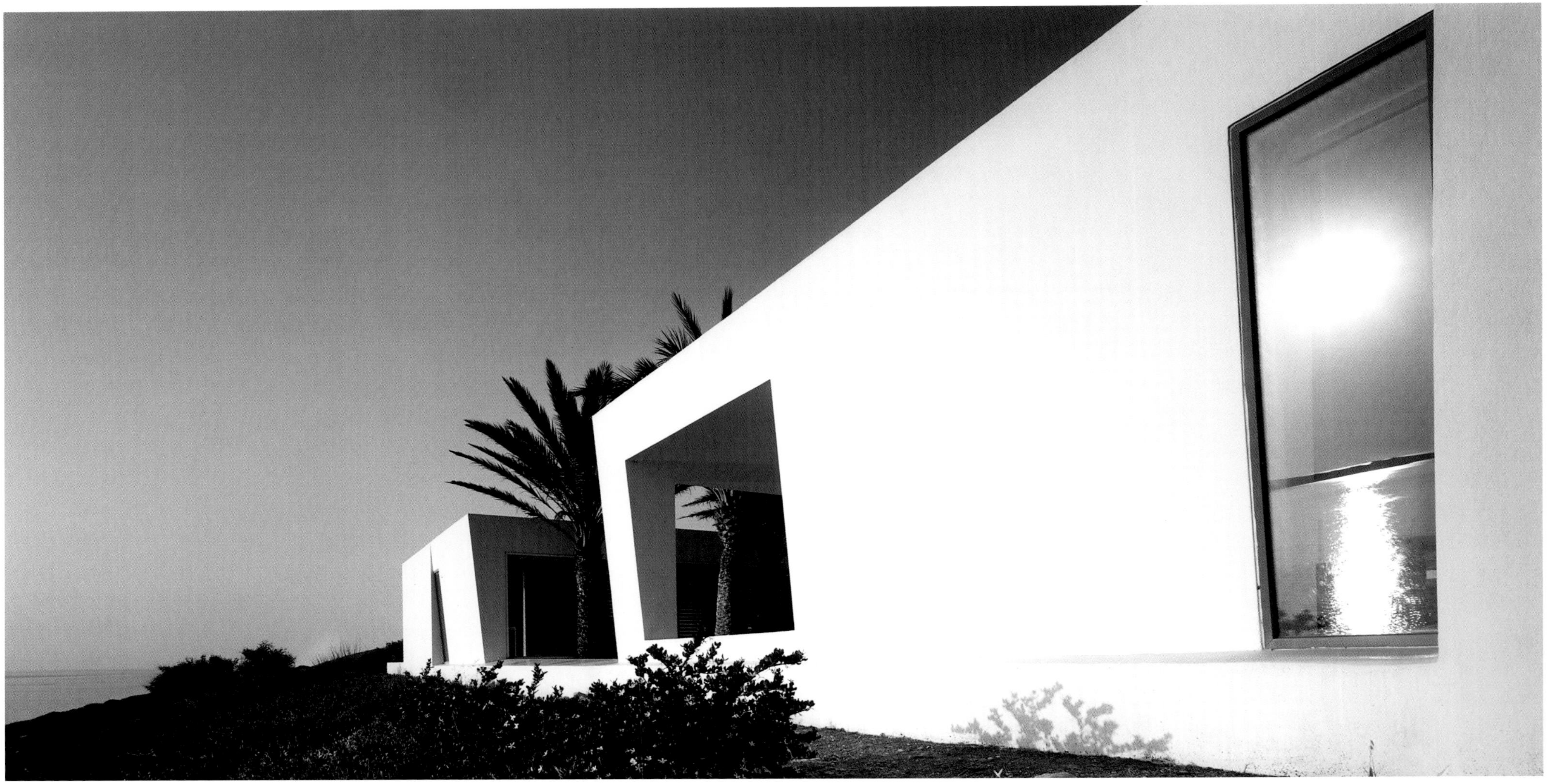

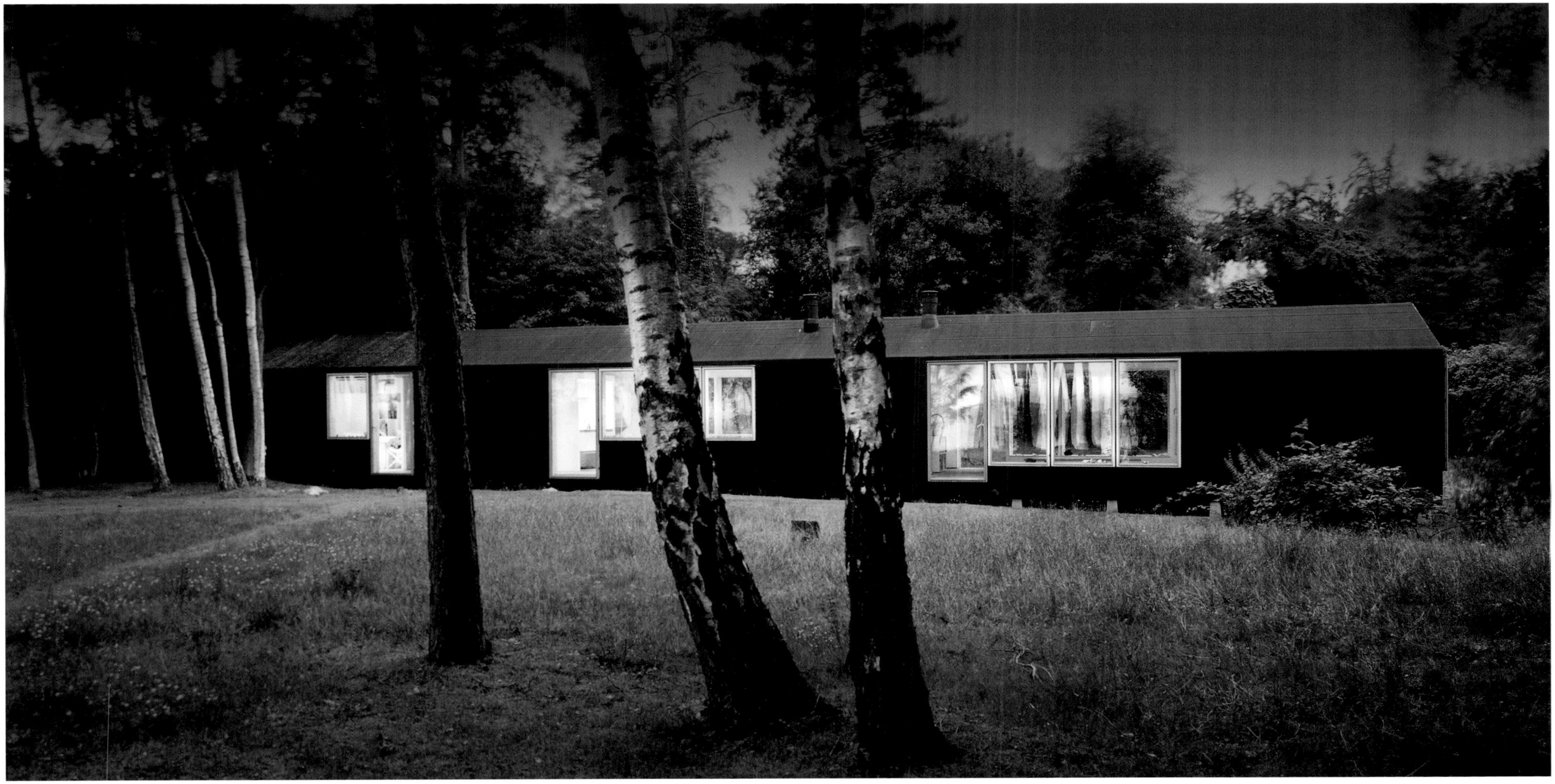

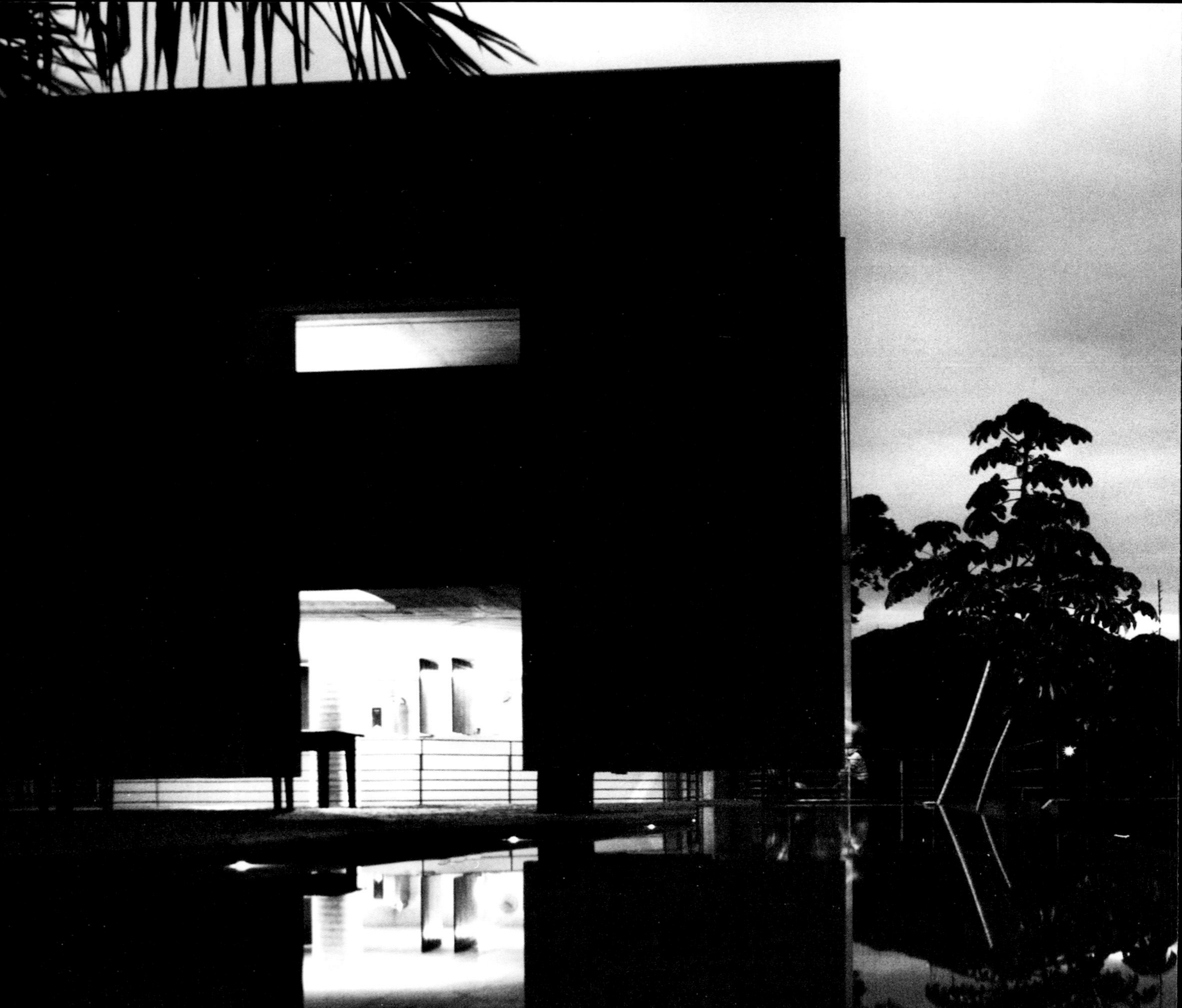

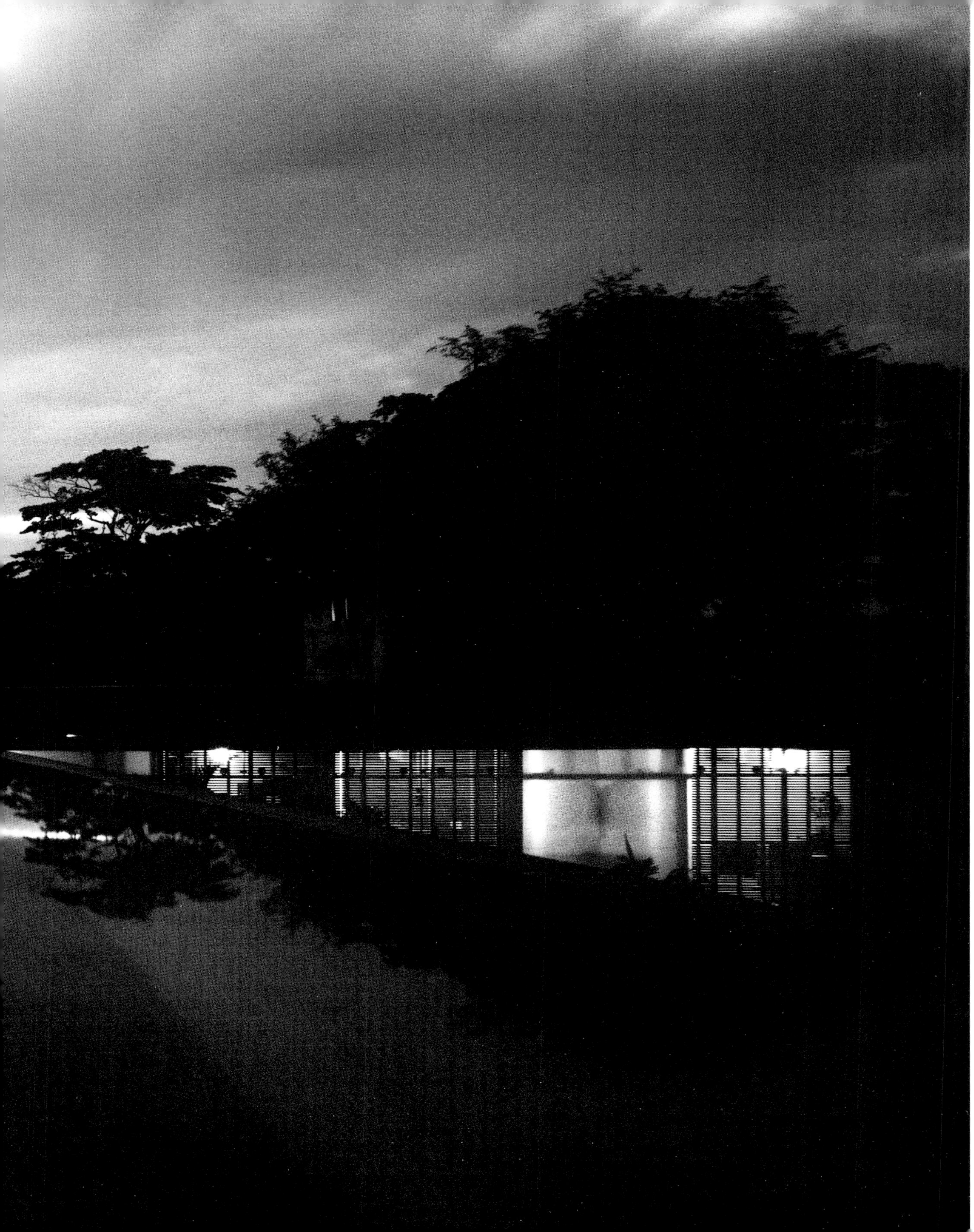

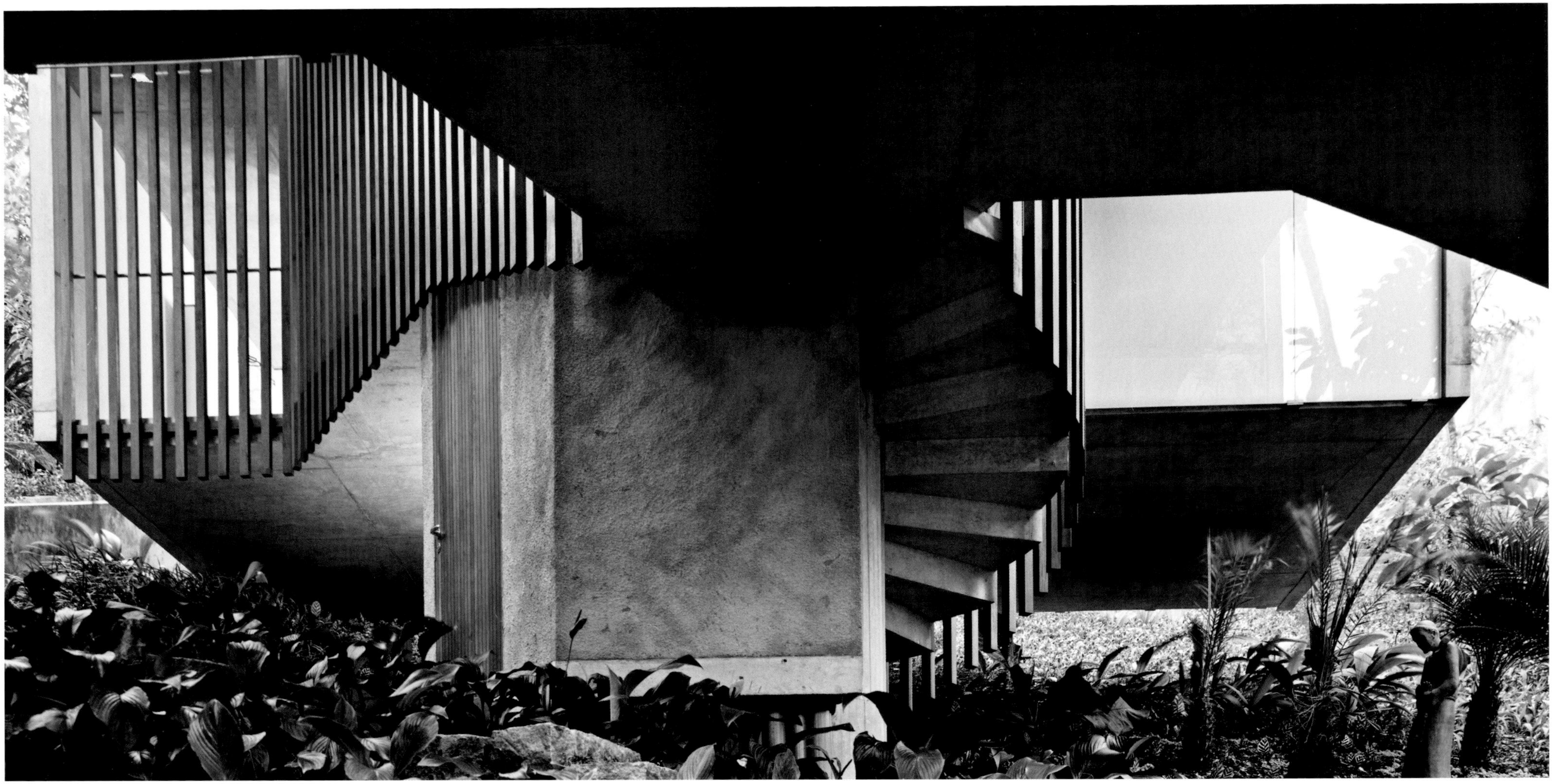

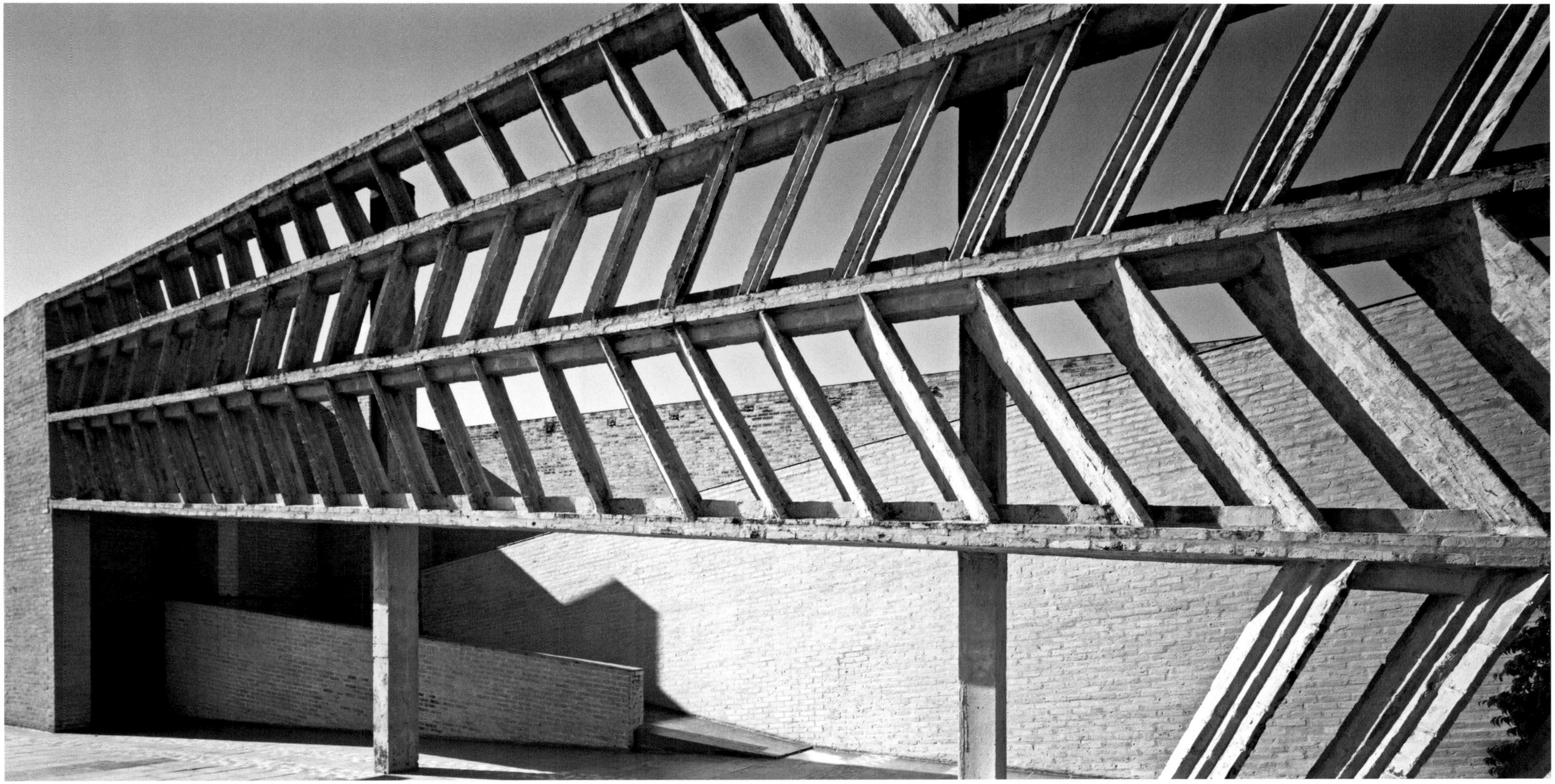

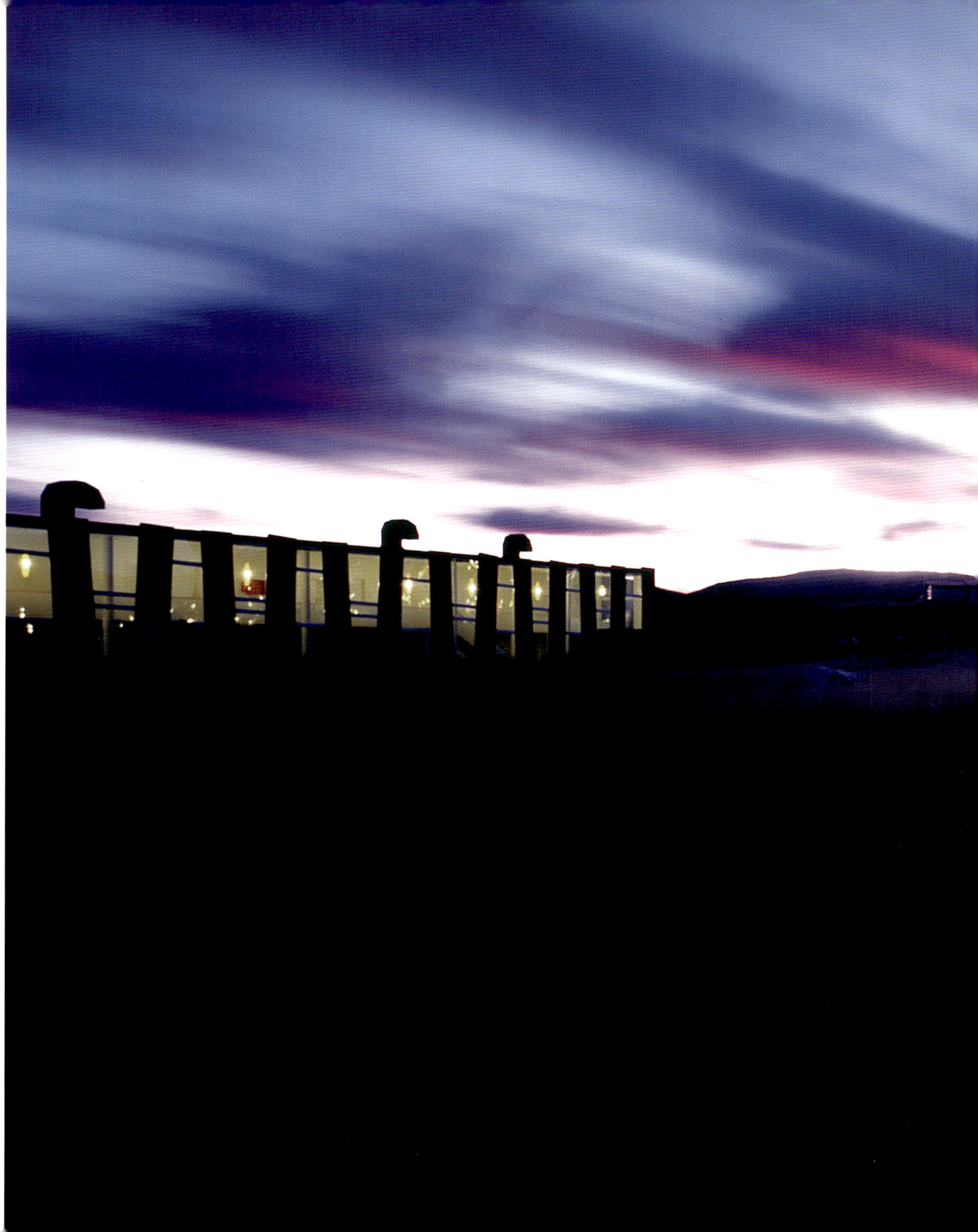

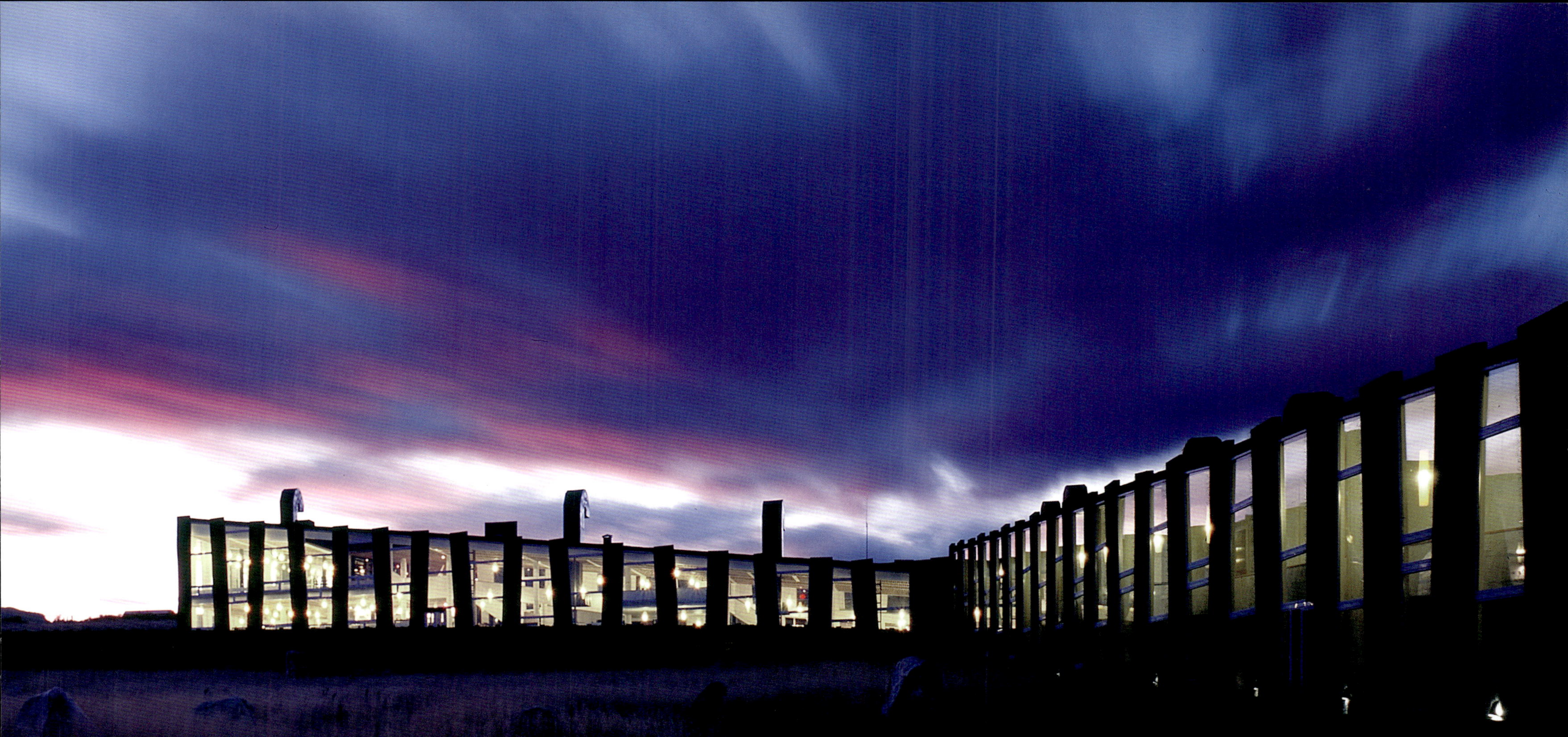

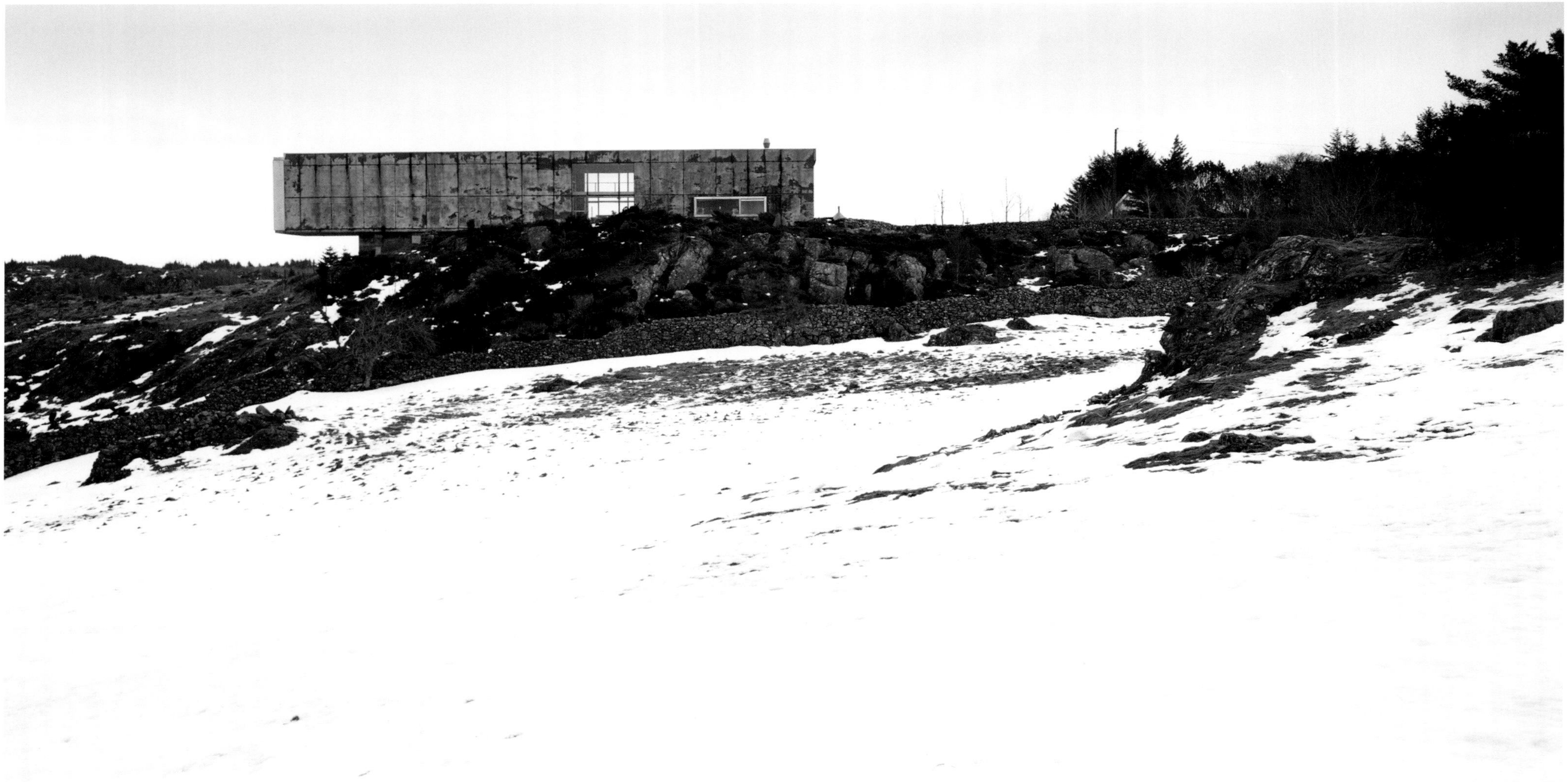

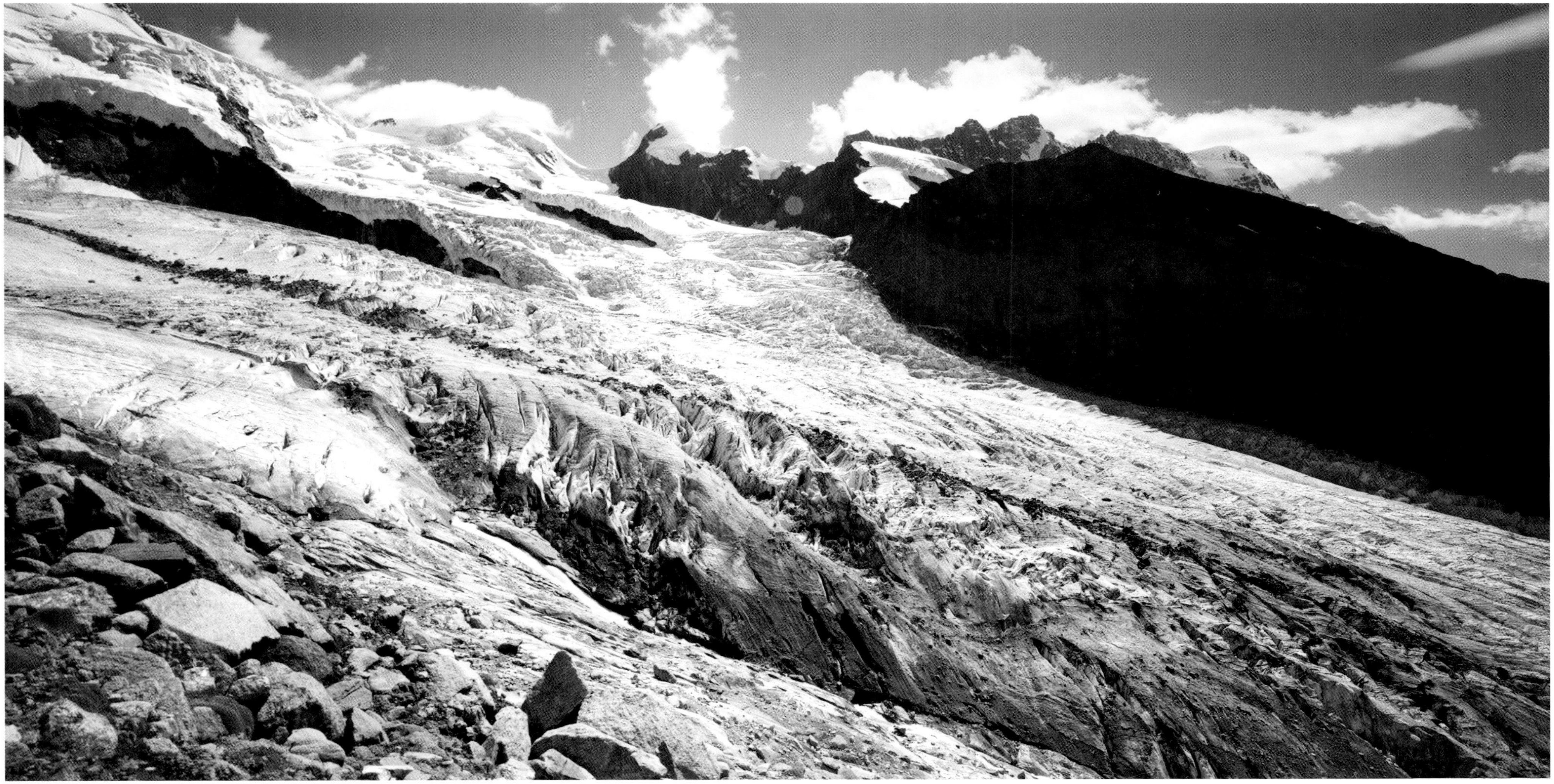

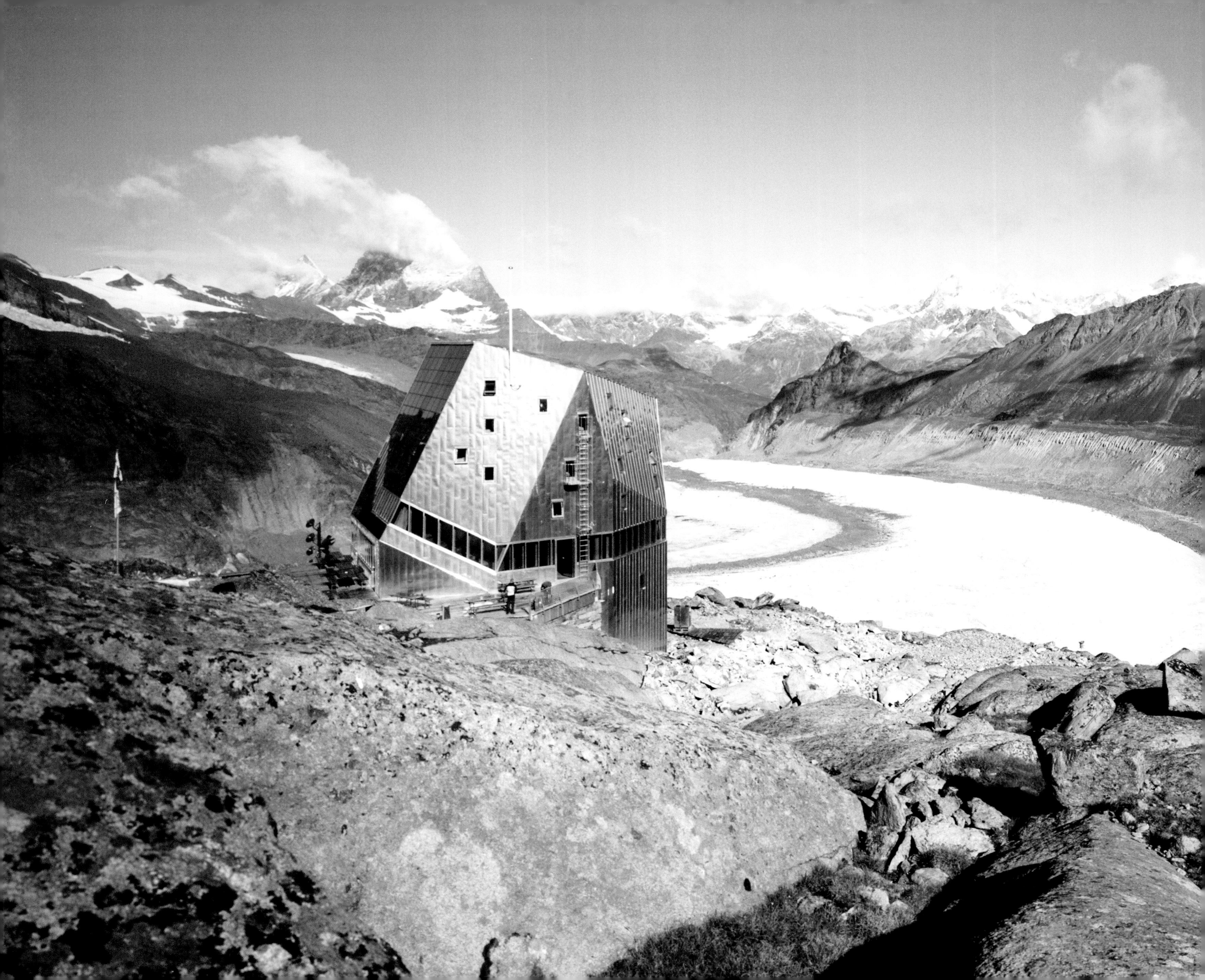

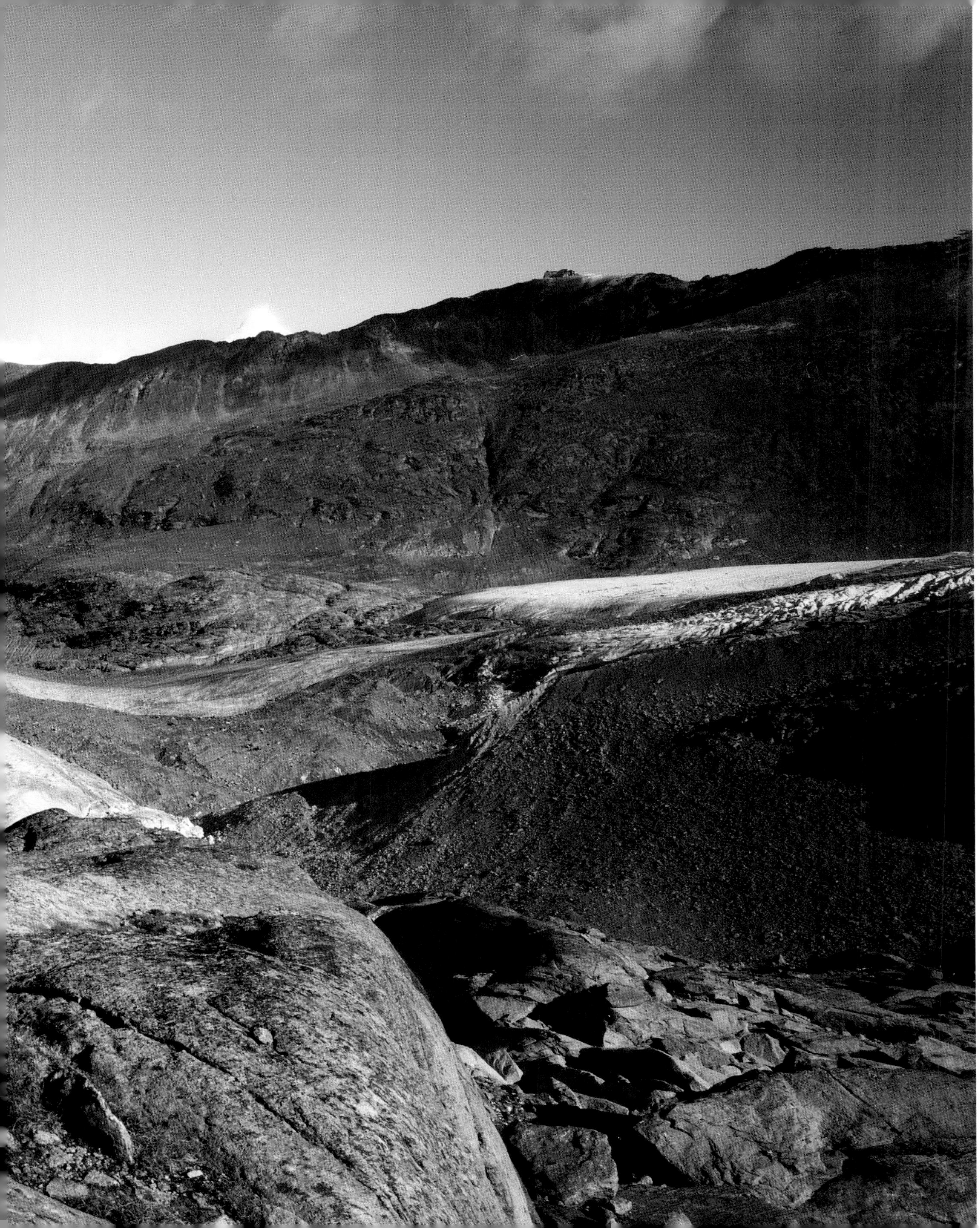

DIMITRIS PHILIPPIDES

In Extremis

ARCHITECTURAL PHOTOGRAPHY in the twentieth century was haunted by Bill Hedrich's paradigmatic 1937 shot of Frank Lloyd Wright's Fallingwater: architecture imposed on pristine nature as a triumph of man's will. But this was soon to change, as it was realized that architecture would gain enormously when shown in its true colors as a human-scale intervention embedded in the vastness of nature. This wasn't only a matter of proportional relation in a competition for attention but rather an issue of overall attitude: landscape had become the real protagonist.

As such, landscape sets its own rules for architectural encroachment. A fully aware architect may now draw inspiration from this imposing, majestic relationship and translate it into a human tribute to the landscape like a religious offering. The photographer, if alerted to such intimate relationships, would then advance his own rendition in harmony with the interaction of these two agents: nature as the protagonist and the architect as an interpreter.

Architectural landscapes eventually began to reflect new concepts of nature and man's response to it. The quick depletion of genuine natural settings due to the overcrowding of the planet triggered a search for the hermit's paradise set further out in the wilderness, while new technological advances permitted the colonization of previously uninhabitable areas. The comic book hero Superman's glacial abode in the Arctic perfectly reflects this notion. More significantly and perhaps more subtly, ideas about urban ecology also affected attitudes toward city life and began to seep into architectural design. Architects were called upon to furnish a boundless habitat, equivalent to *natura naturans*, seemingly based on the art of immateriality. If harsh climatic conditions dictated the use of compact, defensible built mass, the opposite would hold in vulnerable, translucent objects set in the urban milieu. As the "new nature," this artificial environment within grasp of contemporary man could help create an effect similar to those notions of majestic tranquility springing from nature's contemplation.

An outpouring of new, exciting architecture has thus emerged world-wide to satisfy this widely-felt need. As new paths in design were being explored, the ways of looking at architecture itself, from the past or the present, were equally affected. Nothing was more conspicuous than the influence this approach had on architectural landscape photography. Erieta Attali's work lies squarely within this challenging and hazardous field of constant experimentation and ferment.

Attali unearths and fully exposes the overpowering presence of nature in works of architecture nesting in remote, inhospitable landscapes on the outer limits of the known world. She also manages to disguise other landscapes, relatively tame and more familiar to us, by turning them into similarly ominous and forbidding environments. Her approach entails the exploitation of oblique, exaggerated angles of vision, and extreme lighting conditions—the long shadows created at dawn or the setting of mysterious darkness at dusk. The transformation then of perceptiveness is comp ete: architecture often "floats" or "glimmers" in an artificial void reminiscent of the pitch-black skies encompassed by spaceships.

But her even more distinct accomplishment is the capacity to apply such novel ways of looking at architecture to the contemporary urban environment. This new awareness of nature has directed her to emphasize the fluidity and transparency belying modern architecture, resulting in a search for fractured planes severed from their context, the interplay of abstract details brought out by the incidental passage of light, the multiplication of layers that refuse identification. Nature's involvement is made explicit by reflected trees and greenery on glass panes or latent by the narcissistic multiplication of architectural elements reproduced by the same process, so as to create an outlandish "landscape." This impression is accentuated by controlled tonal reversal: long exposures create dazzling strips of artificial light that helps create a 'surreal' atmosphere.

Such techniques block out the conventional representation of a building as a reminiscence of standard drawings and replace it with a perceptive richness closer to that encountered in a dense forest floor where glimpses of shimmering light penetrate the darkness produced by multiple layers of foliage. Some photographs go even further since they emphasize the penetration or encirclement of architecture by nature, seen either as a framed view or as actual physical contact. Under such circumstances architecture is depicted as a frail, insignificant intruder to the natural setting.

By such a process of X-raying architecture Attali bisects its mass and allows its inner structure, its fragility to emerge. This notion is sometimes further exploited by reducing architecture to a mere strip in the background of a landscape; in this manner the Renaissance custom of allowing a landscape to occupy the distant background of a painting is reversed.

The daring exclusion of man-made structures from a landscape is a further proof of Attali's determination to go beyond architecture, which is no longer a compulsory ingredient in the depiction of an architectural landscape. In such views, architecture is ever-present because these seemingly 'uninhabited' images of nature directly refer to architectural properties: monumentality, abstractness, rhythmic repetition, or uniform texture. With the help of a rigorous set of allusions, architecture thus returns as a grand metaphor. Its presence goes much deeper than the conventional colonization of natural landscape by structures and buildings. Attali's keen perception can foresee this eventual shift of roles and record it convincingly on film.

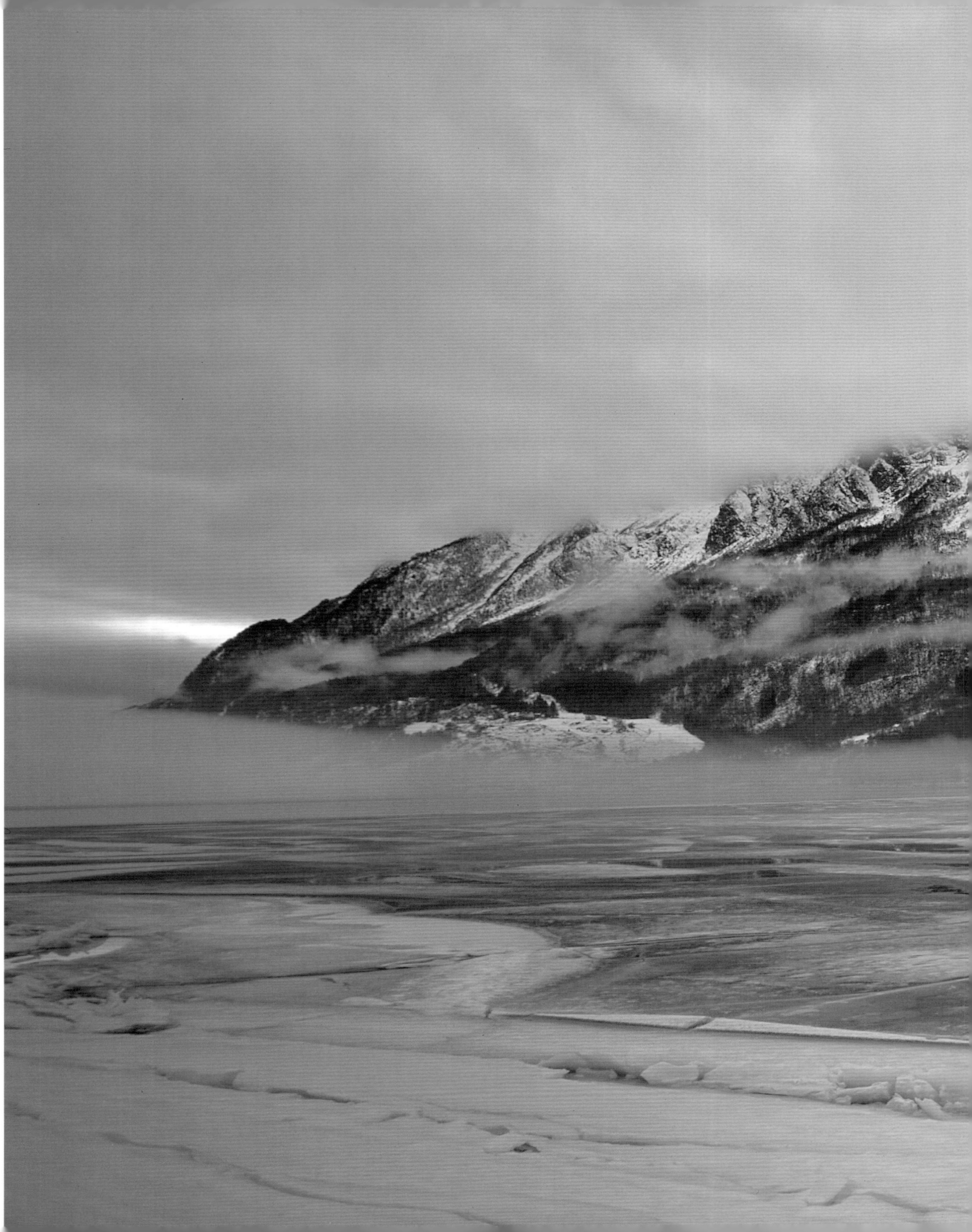

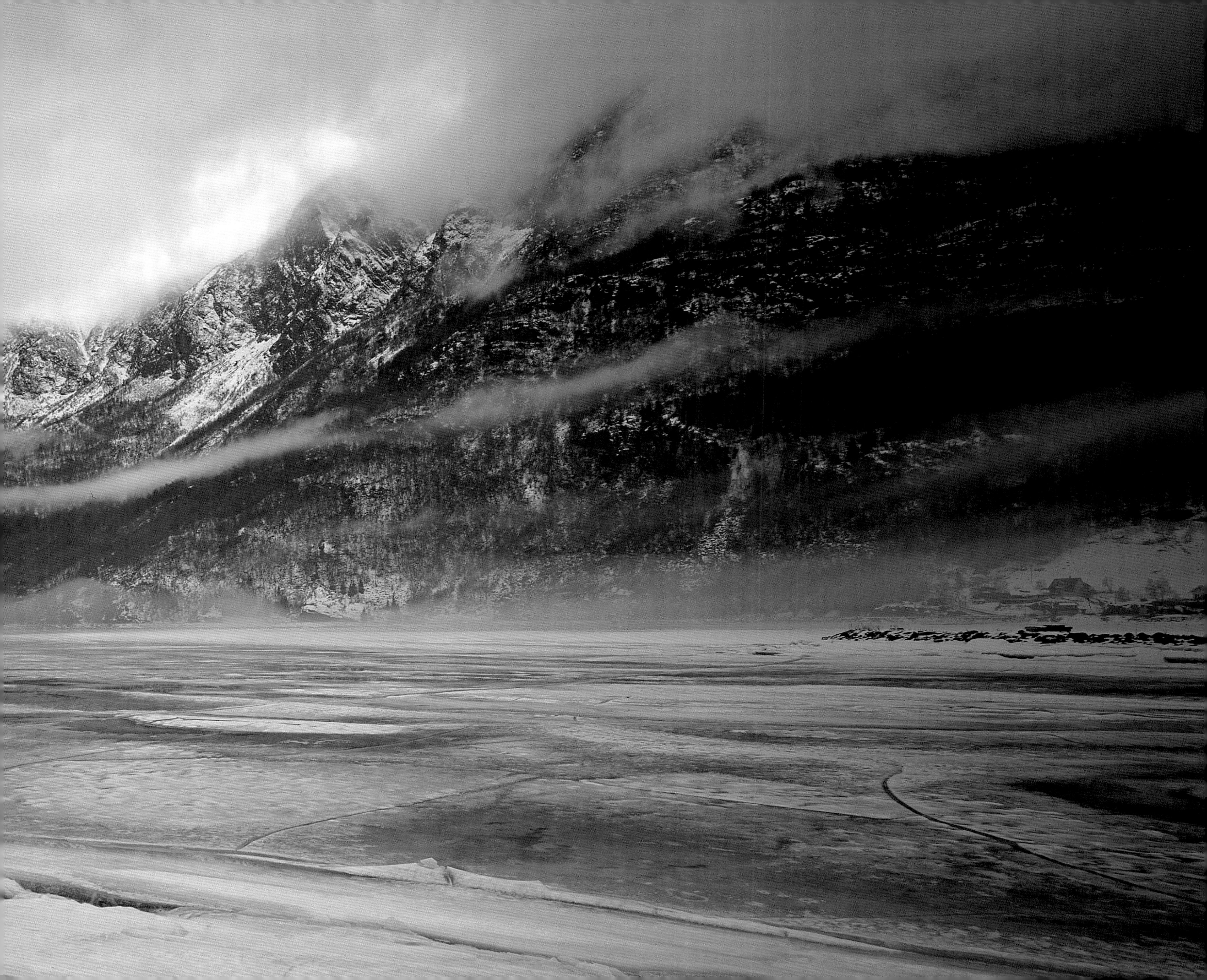

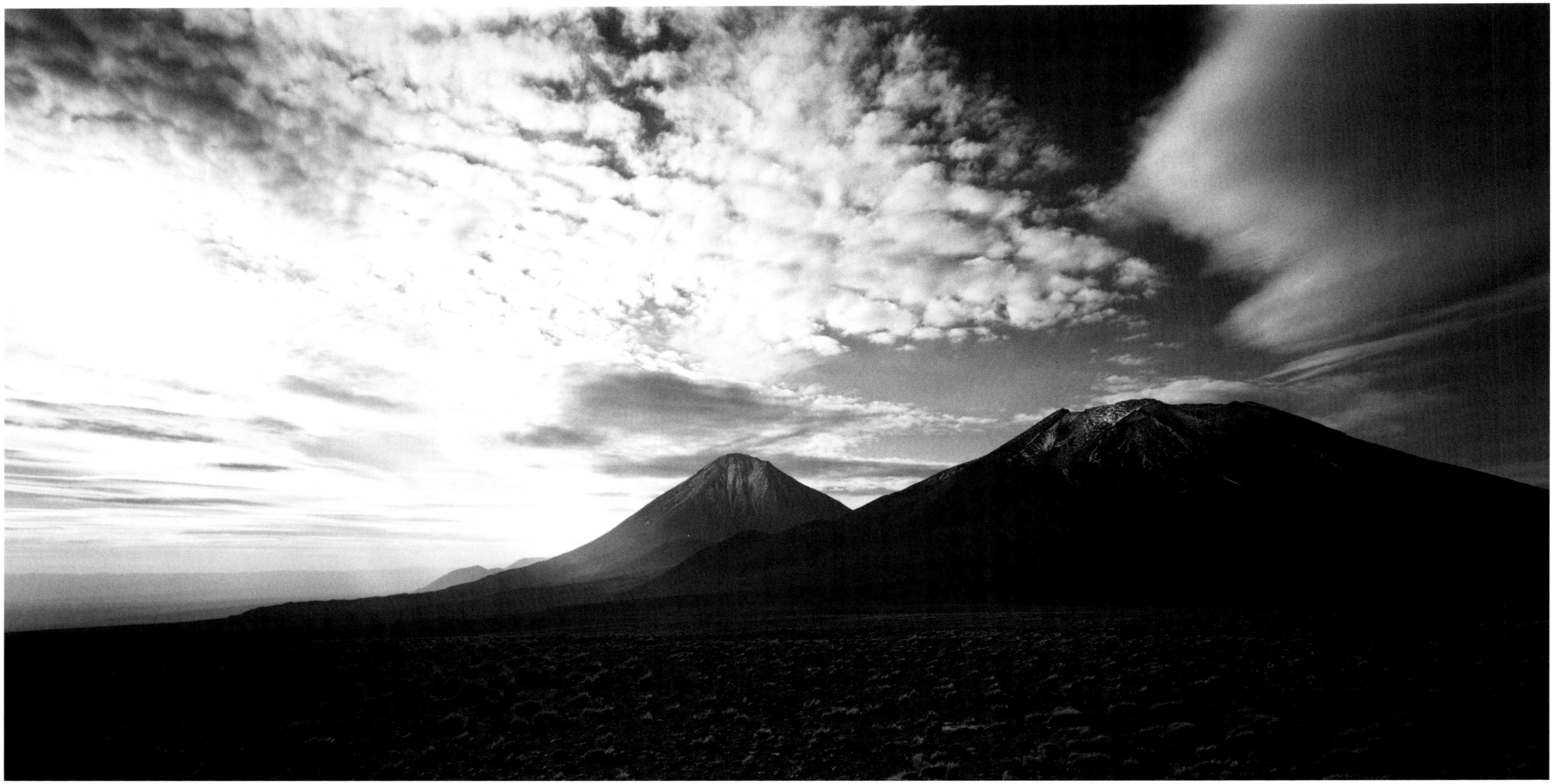

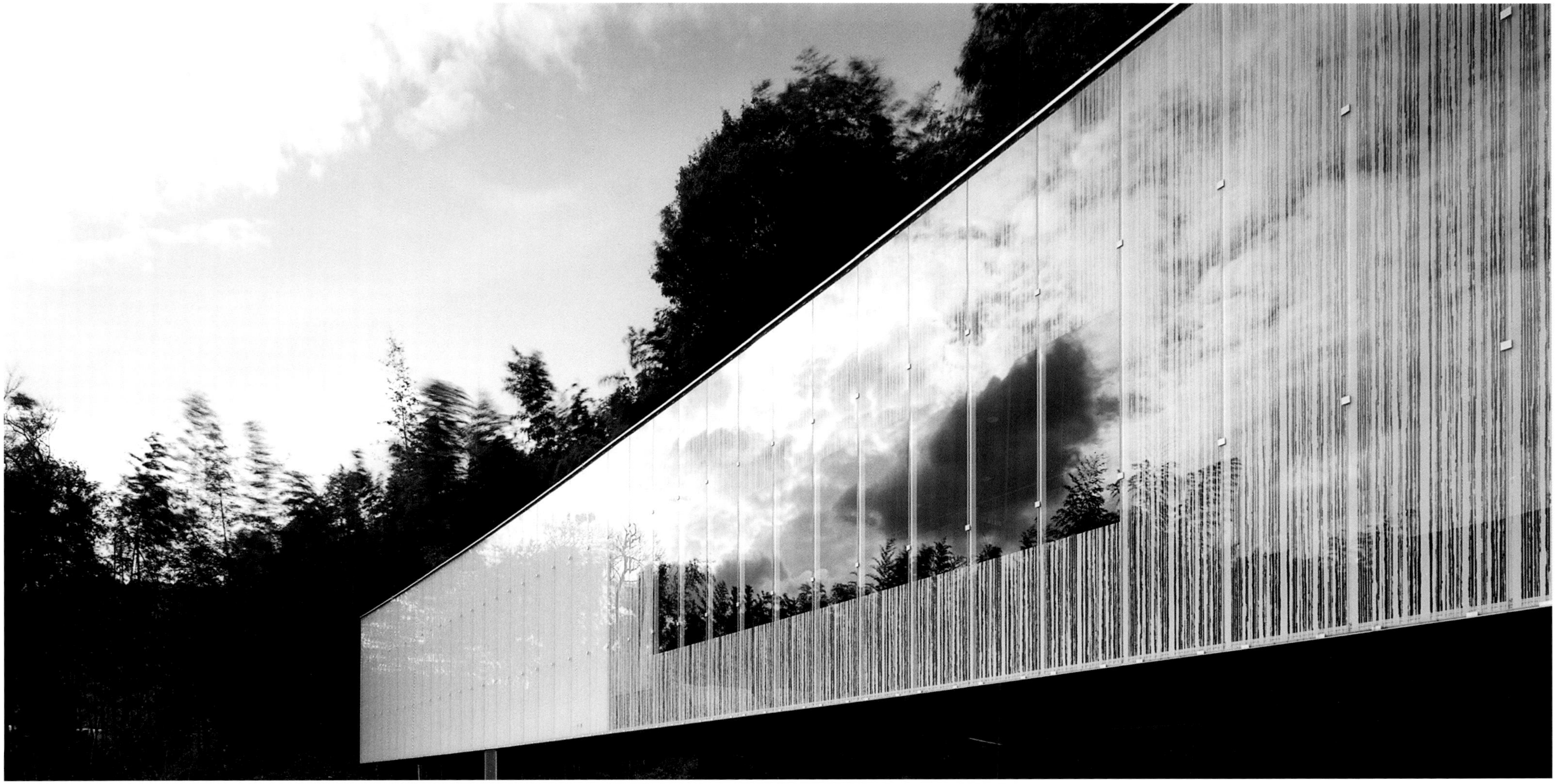

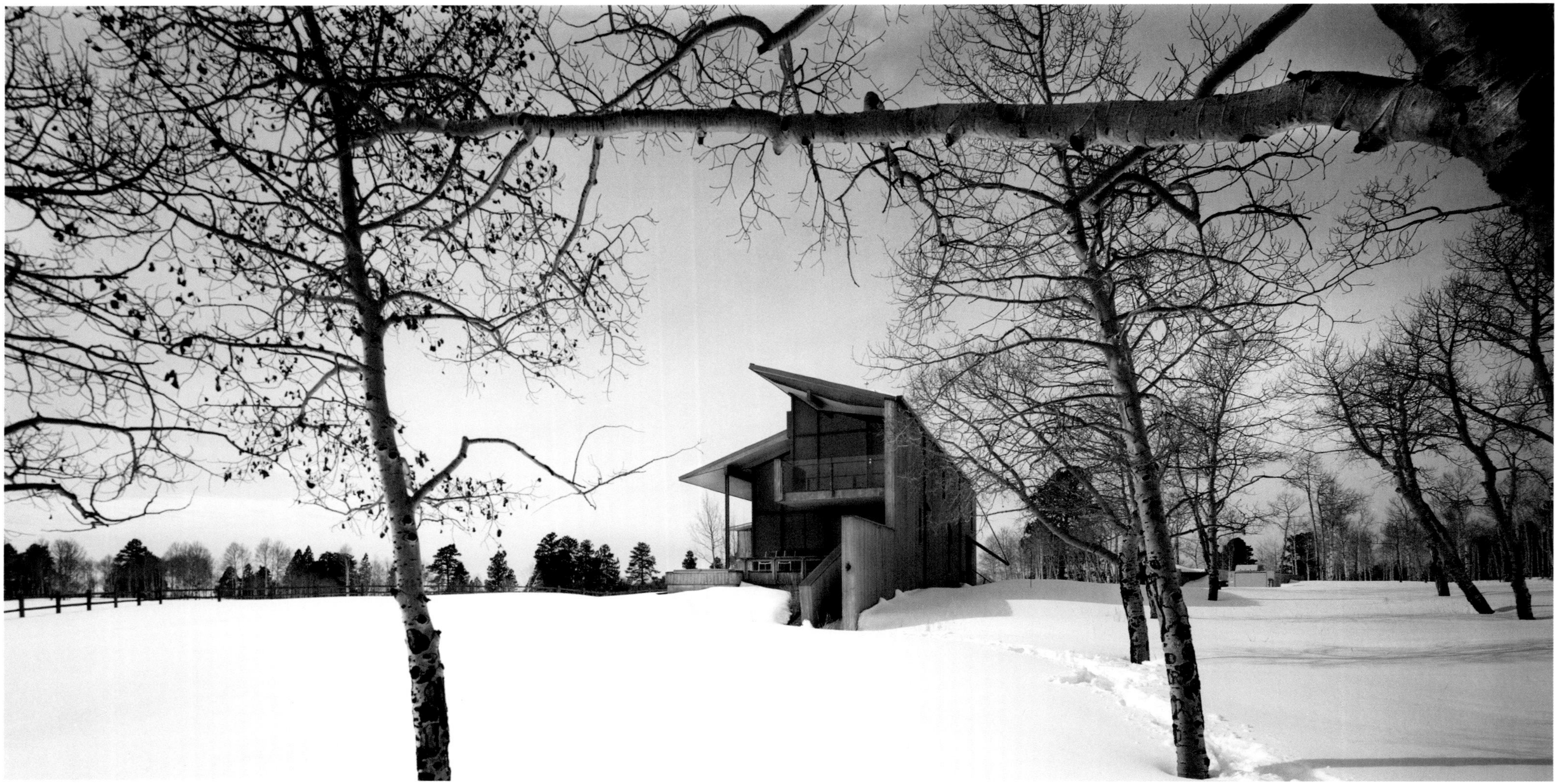

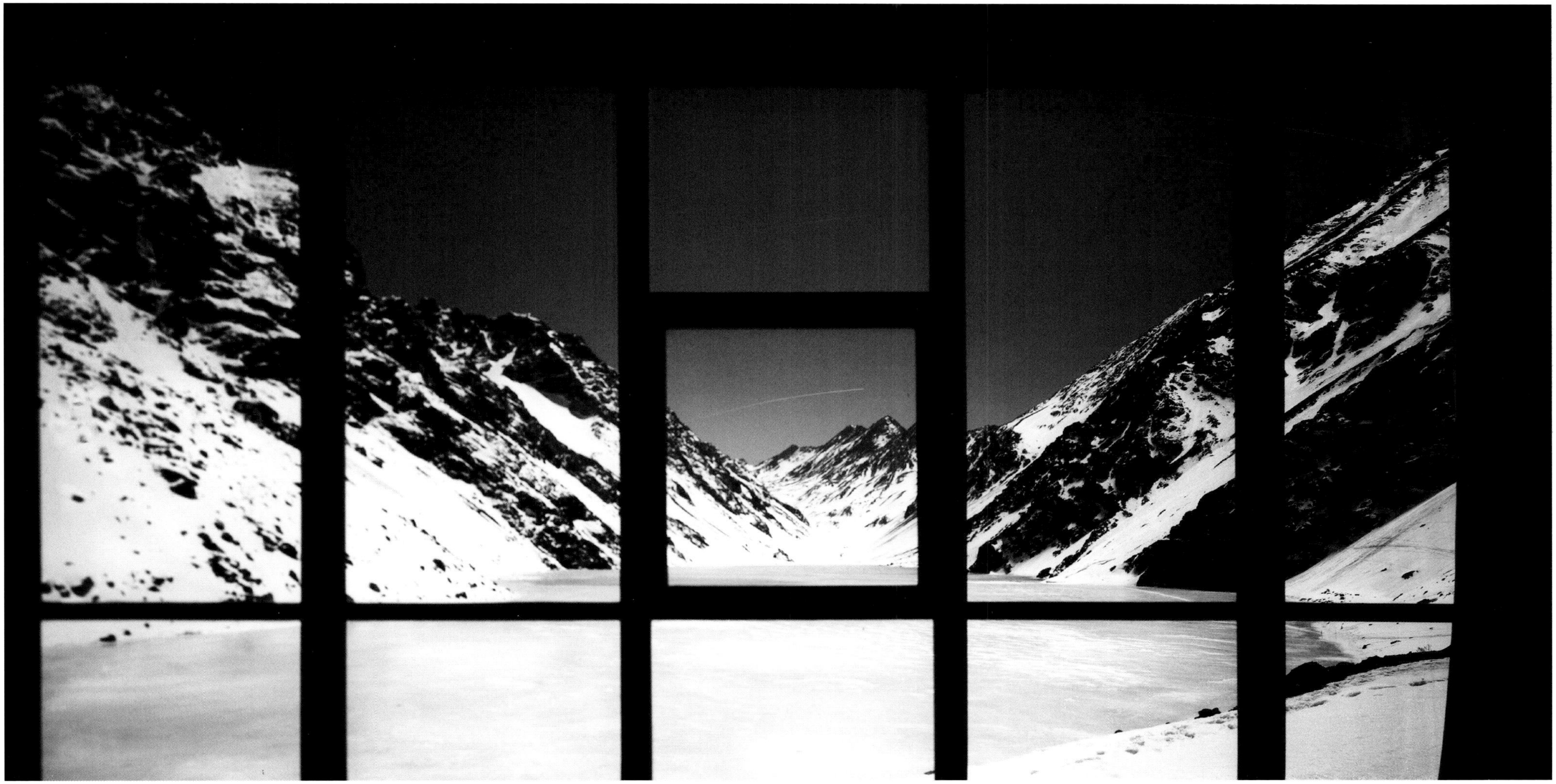

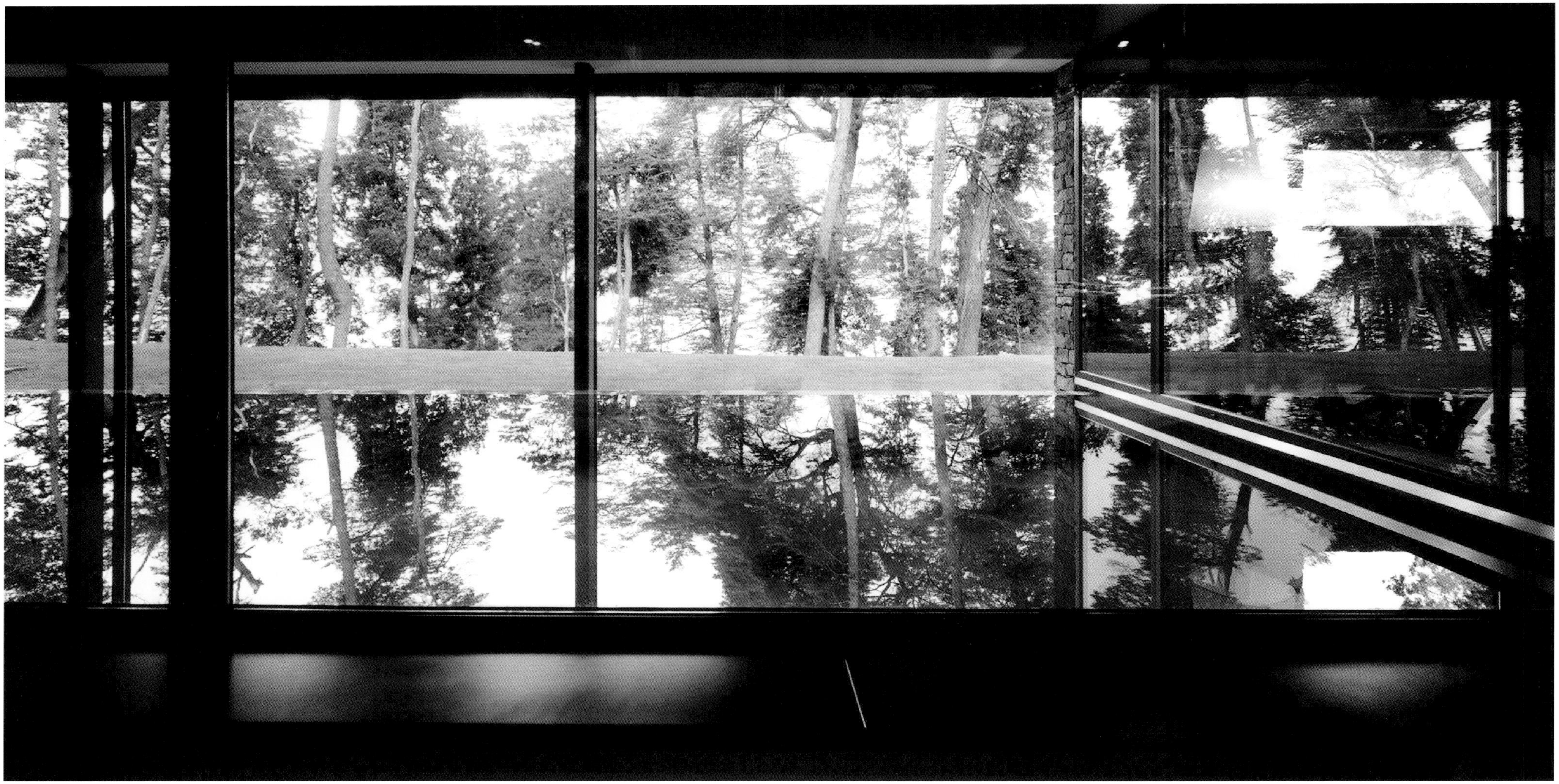

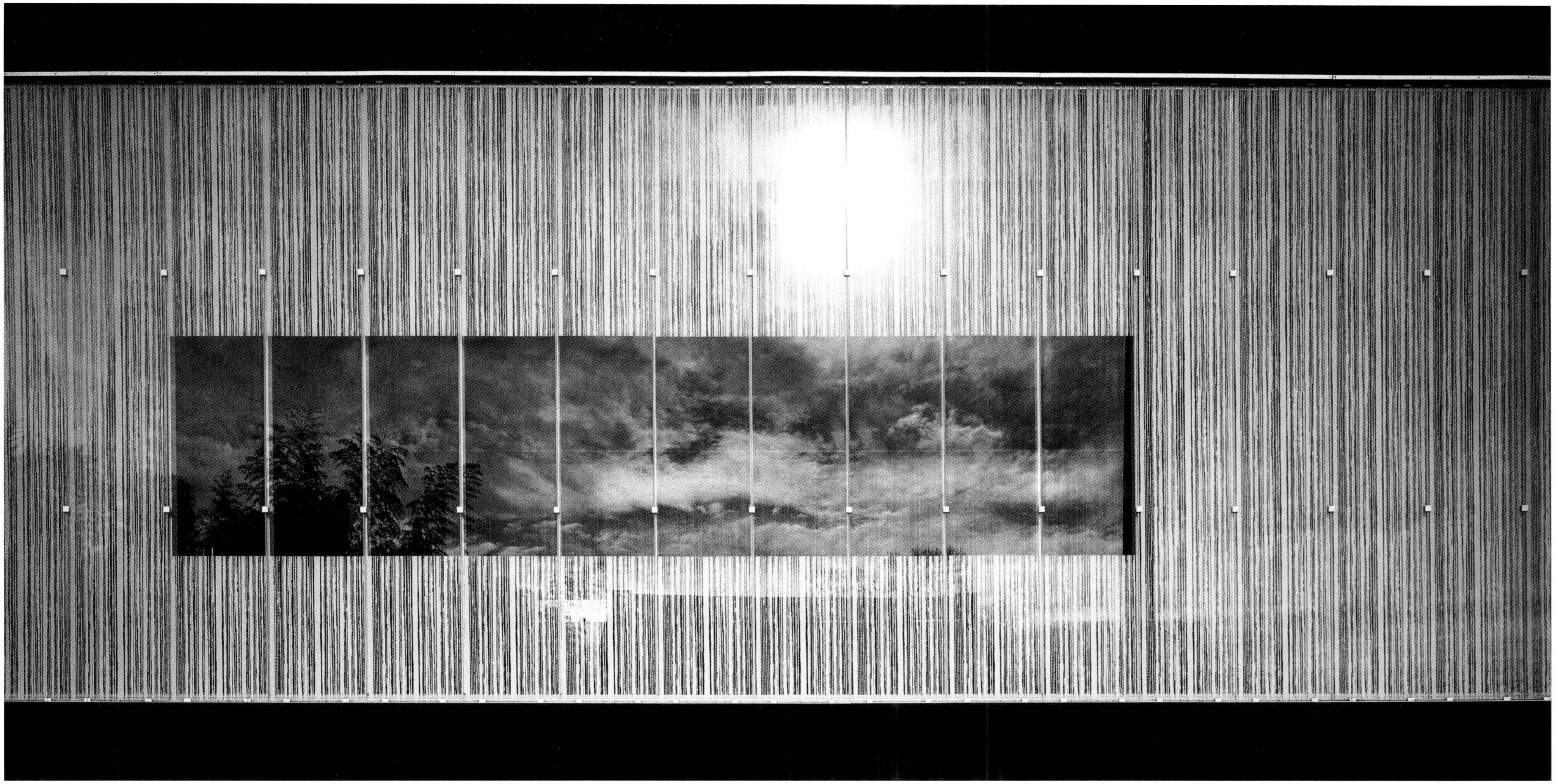

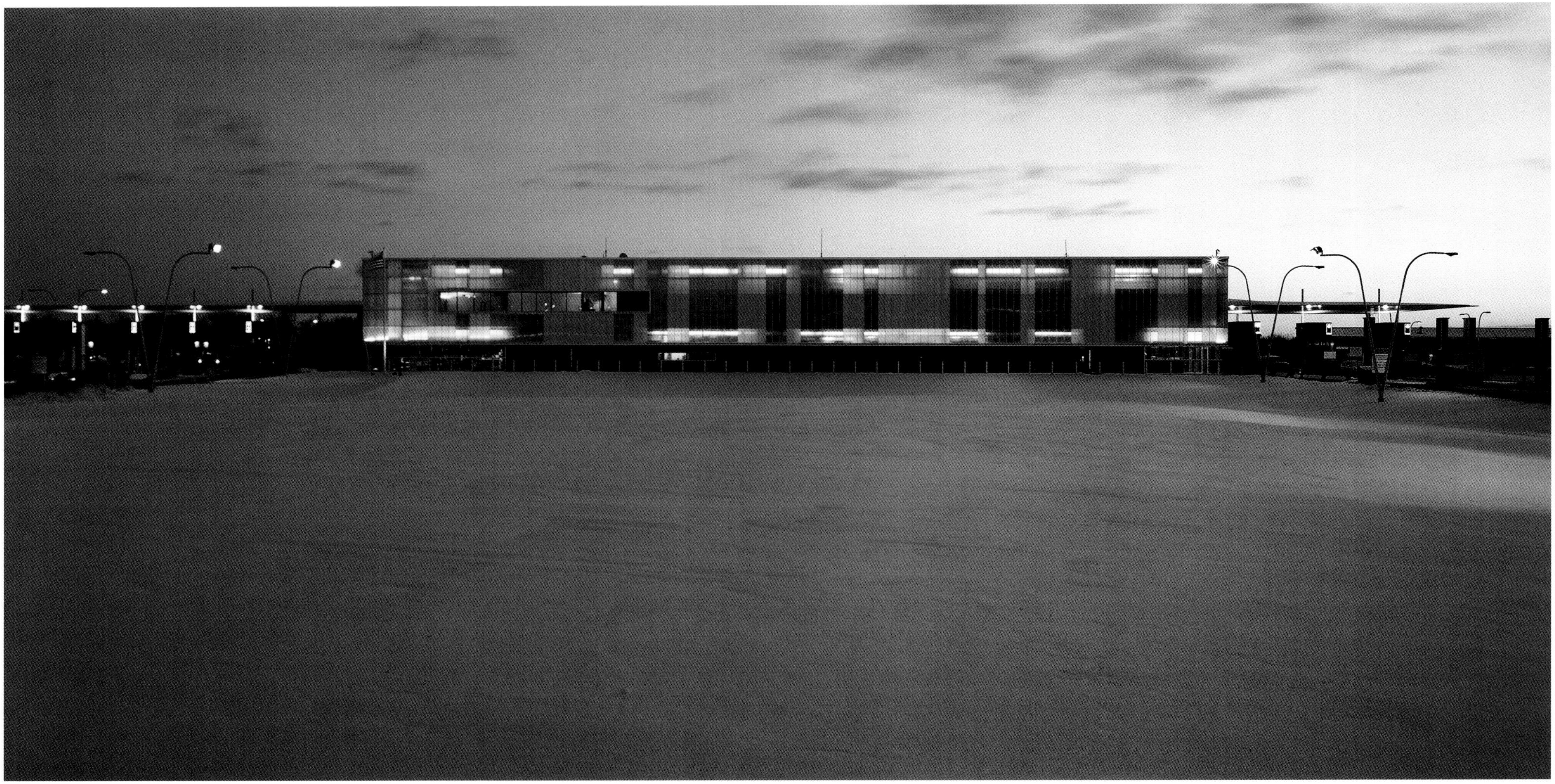

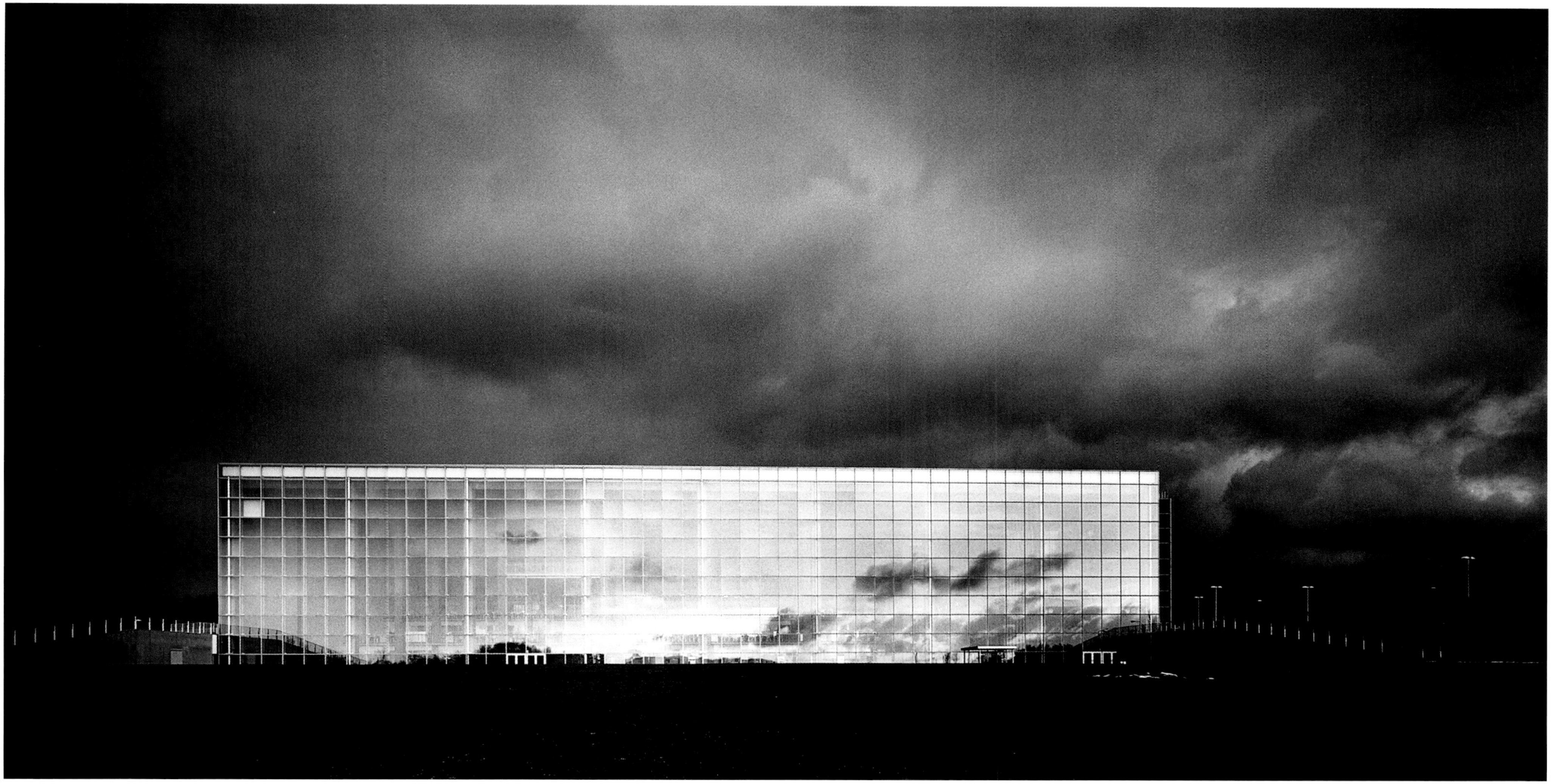

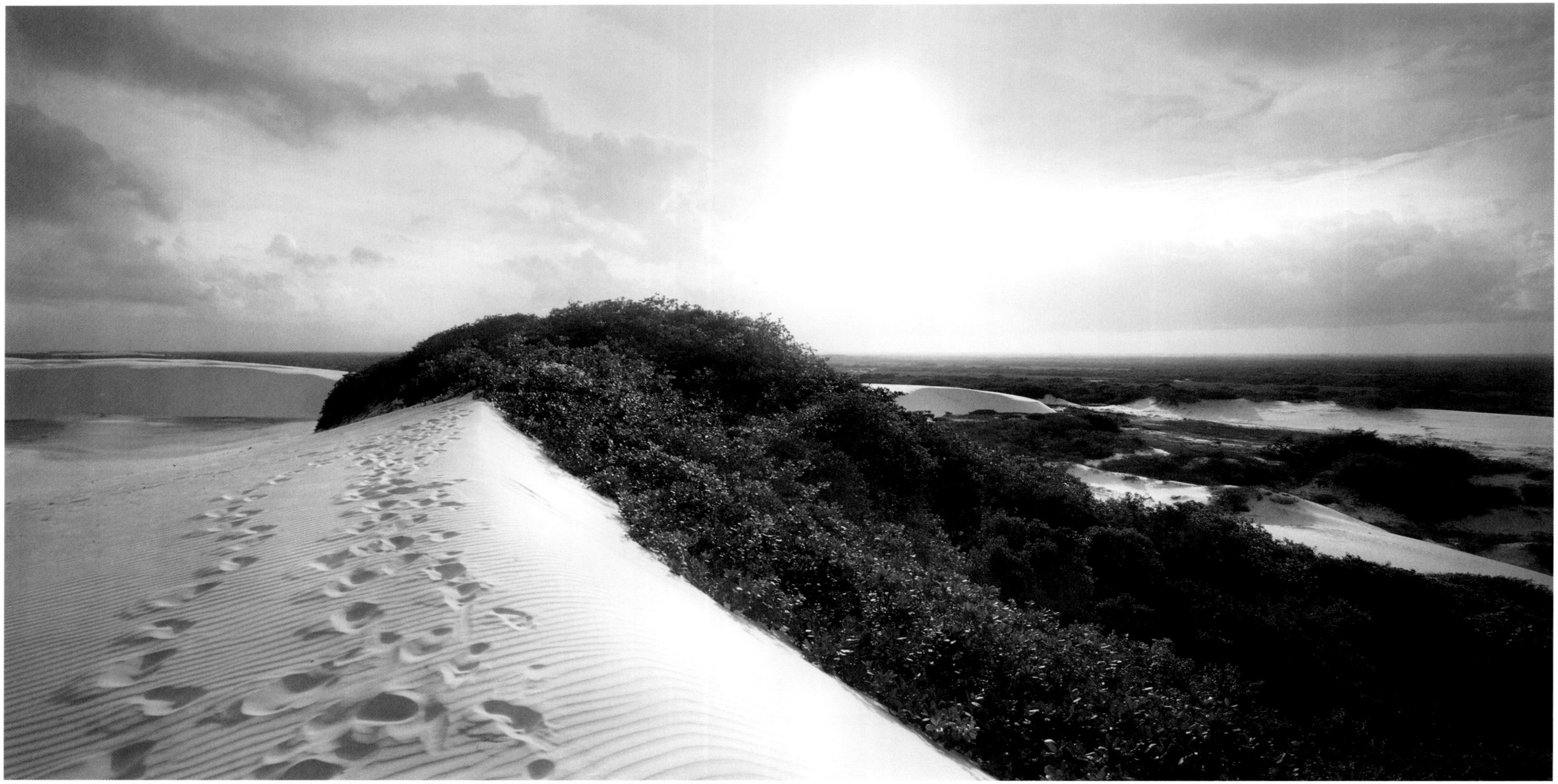

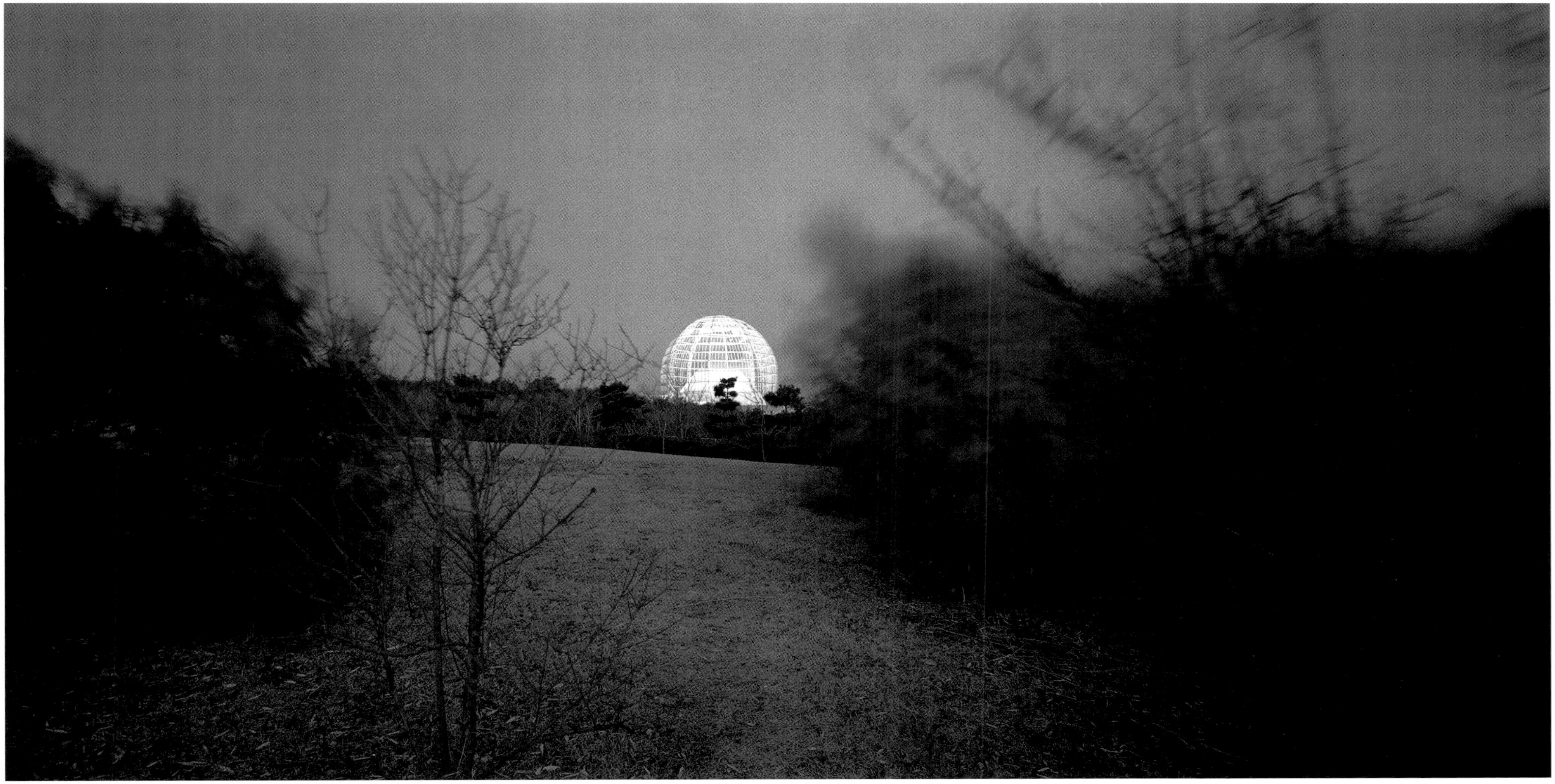

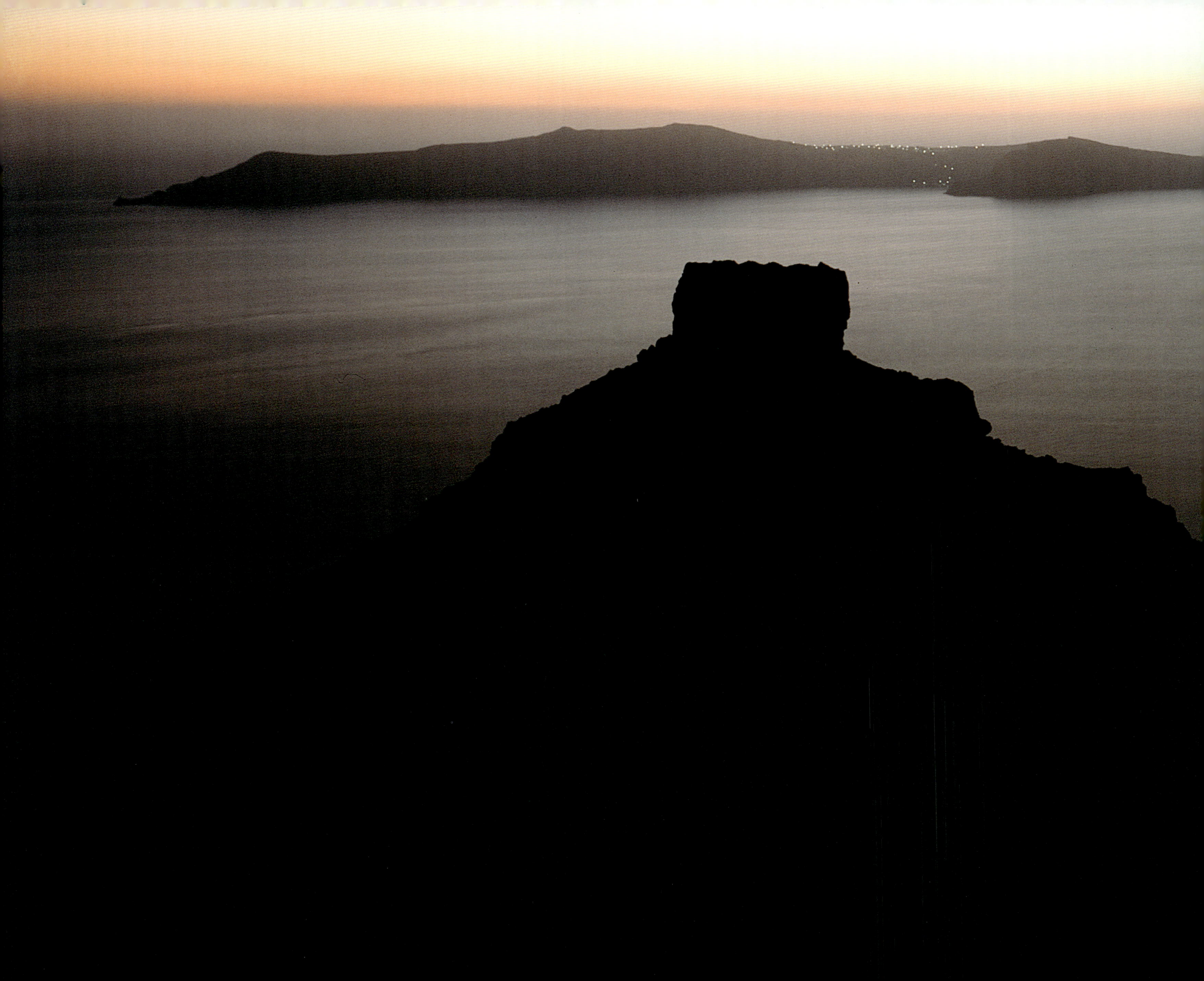

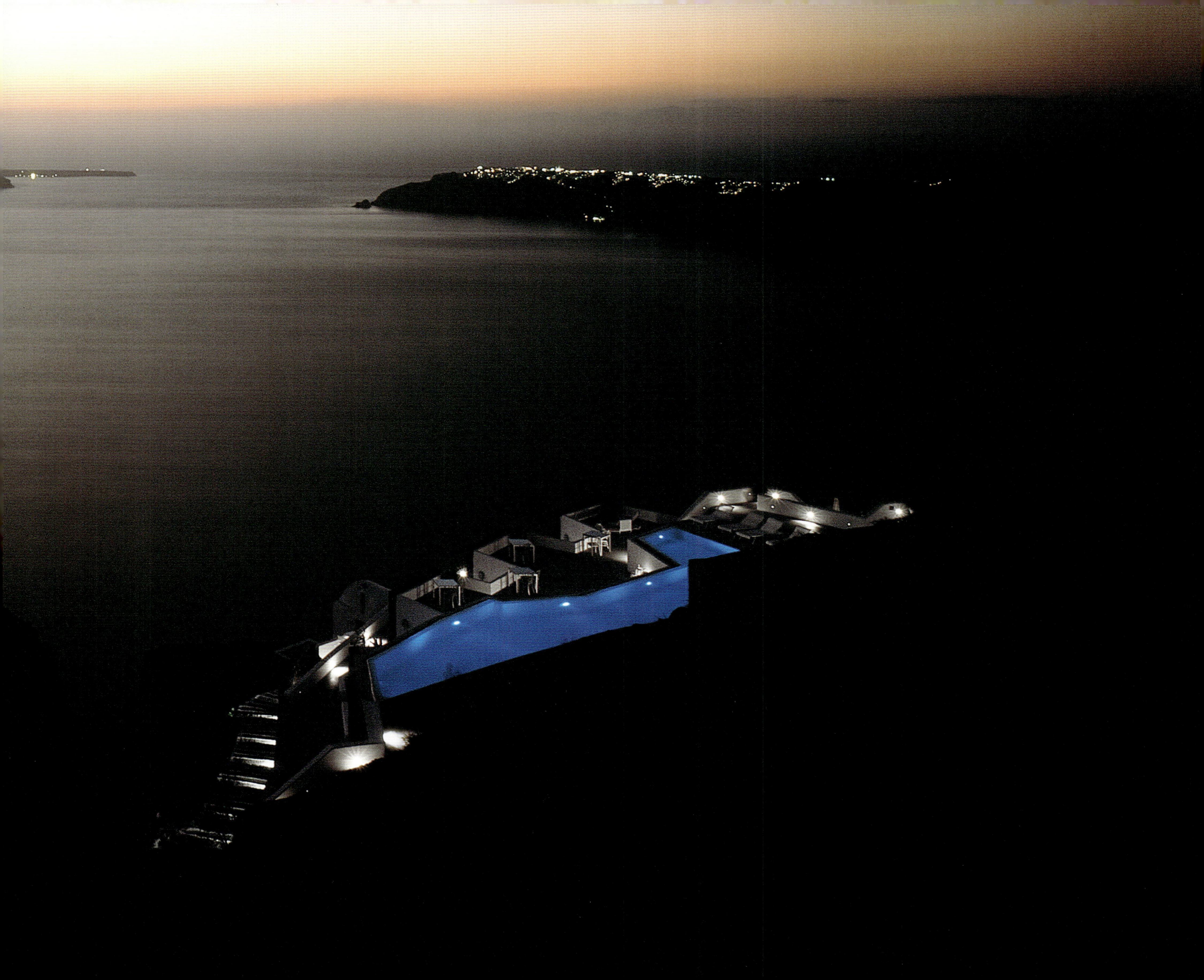

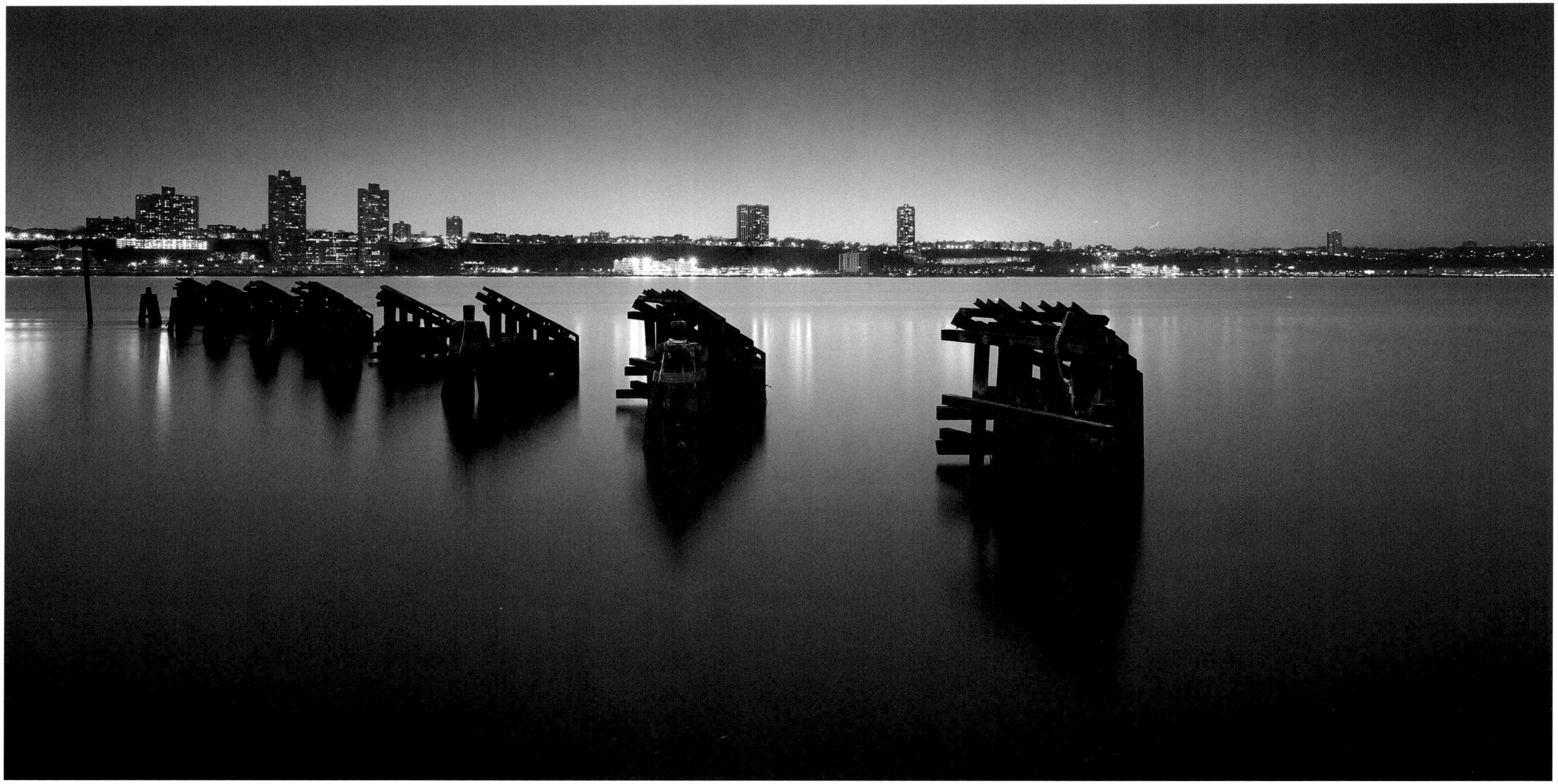

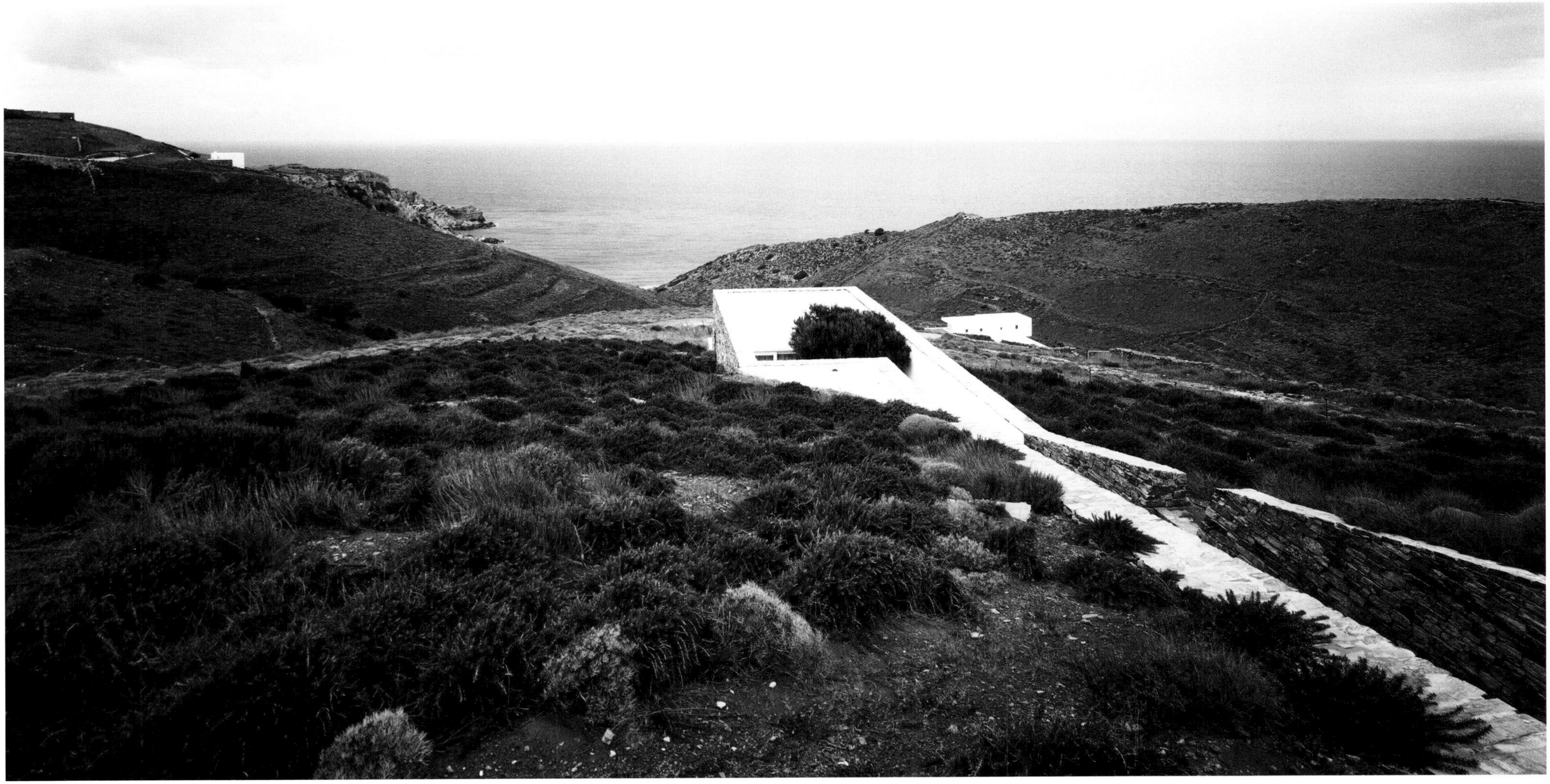

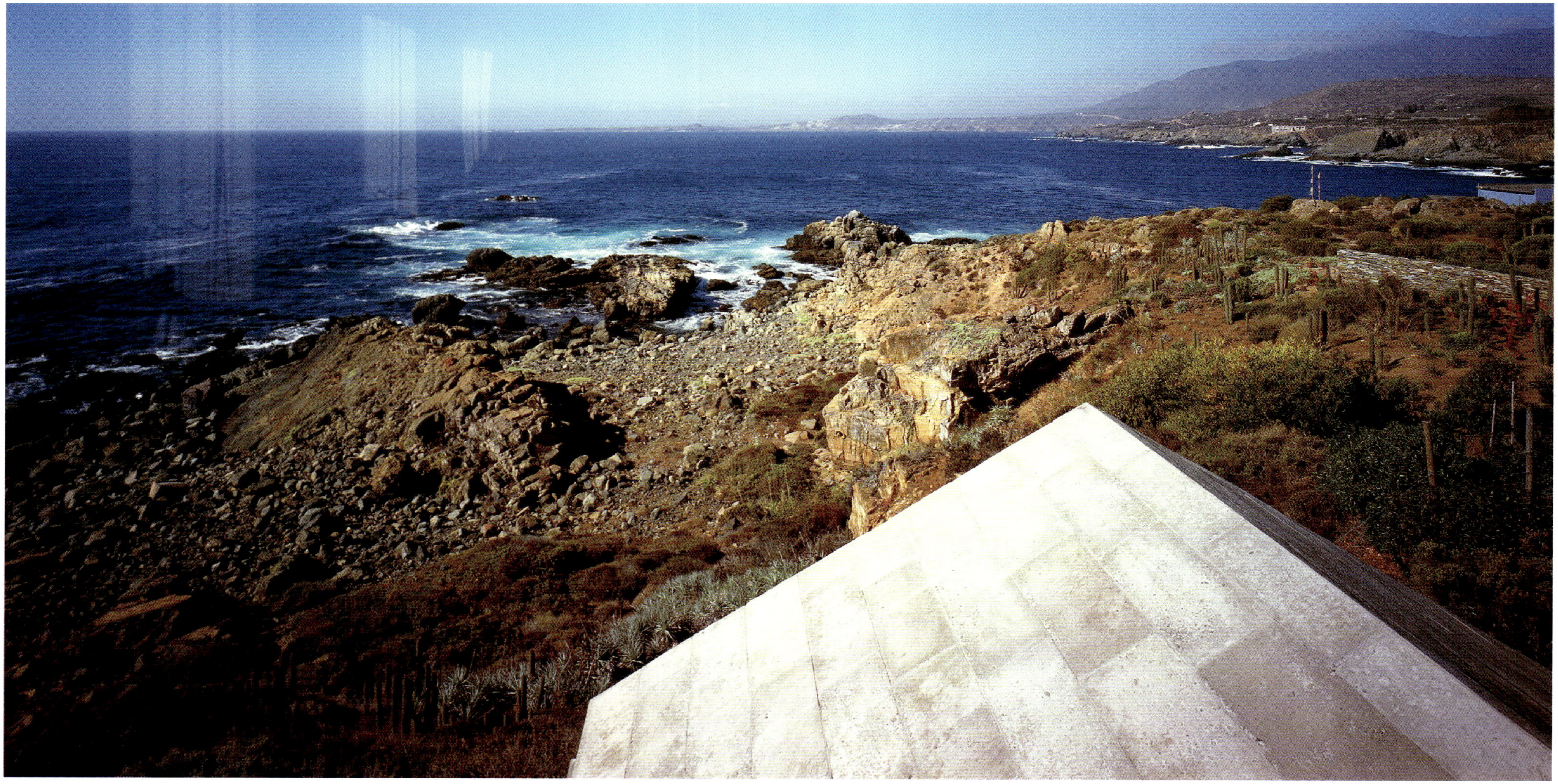

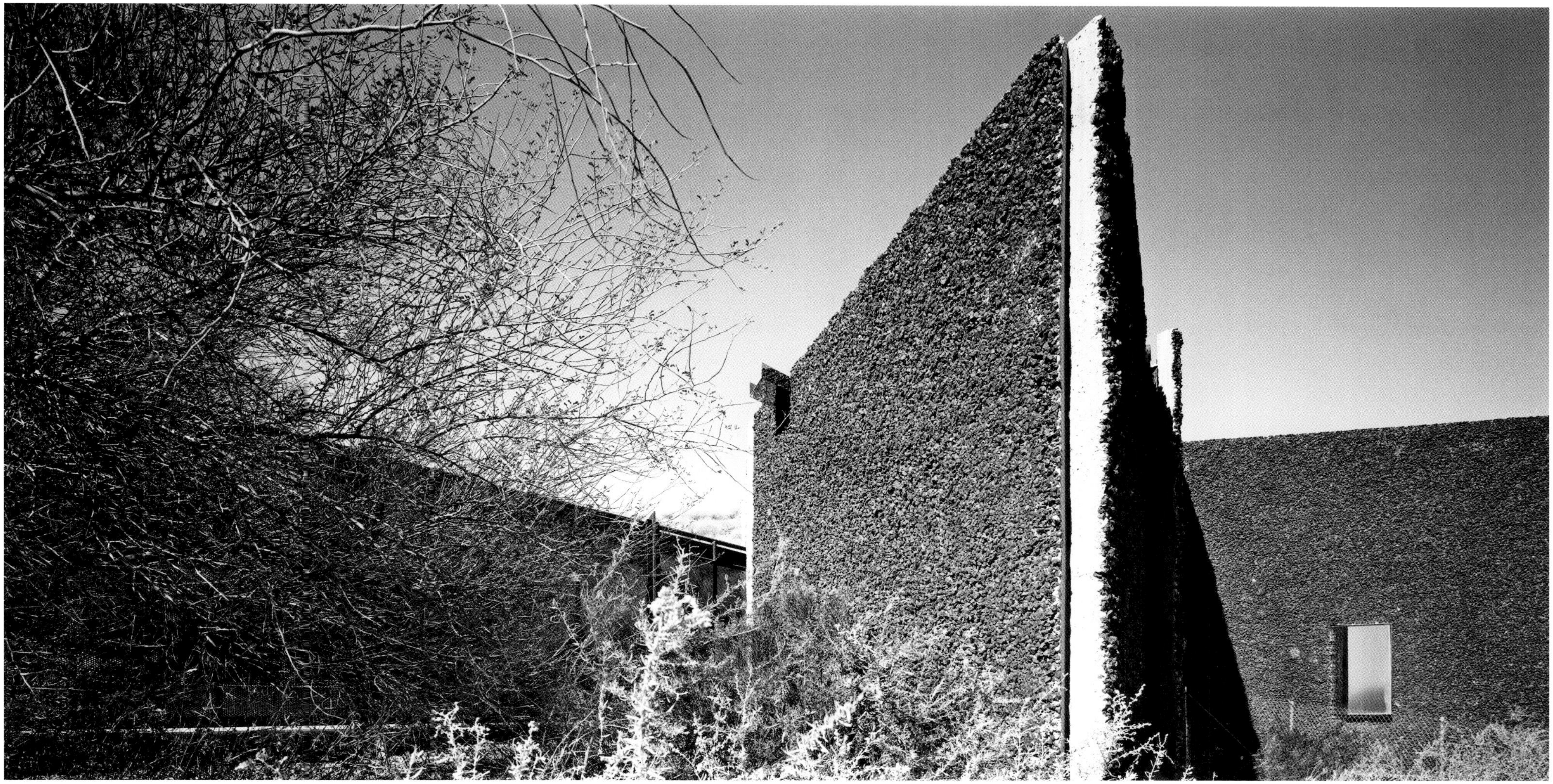

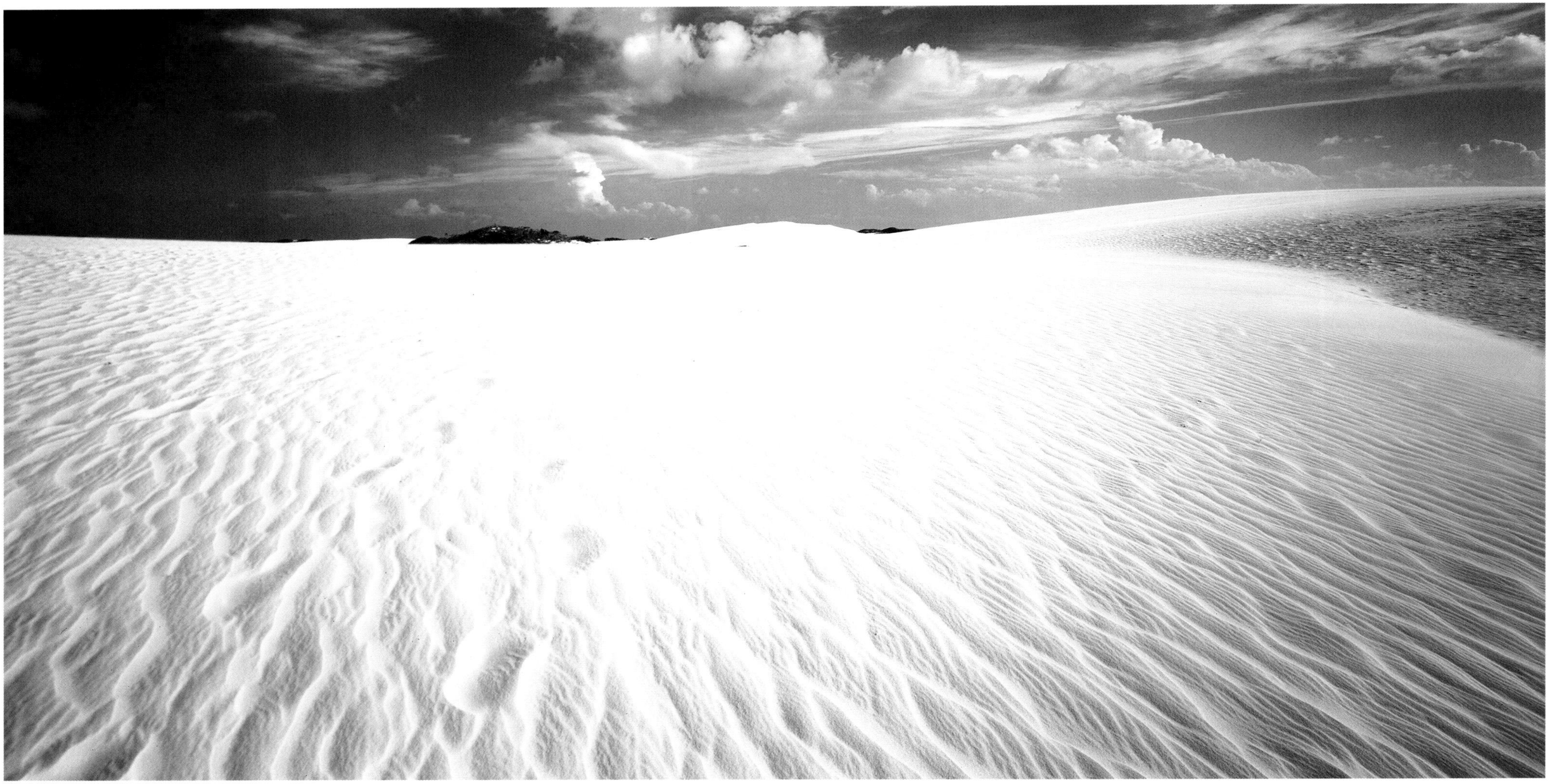

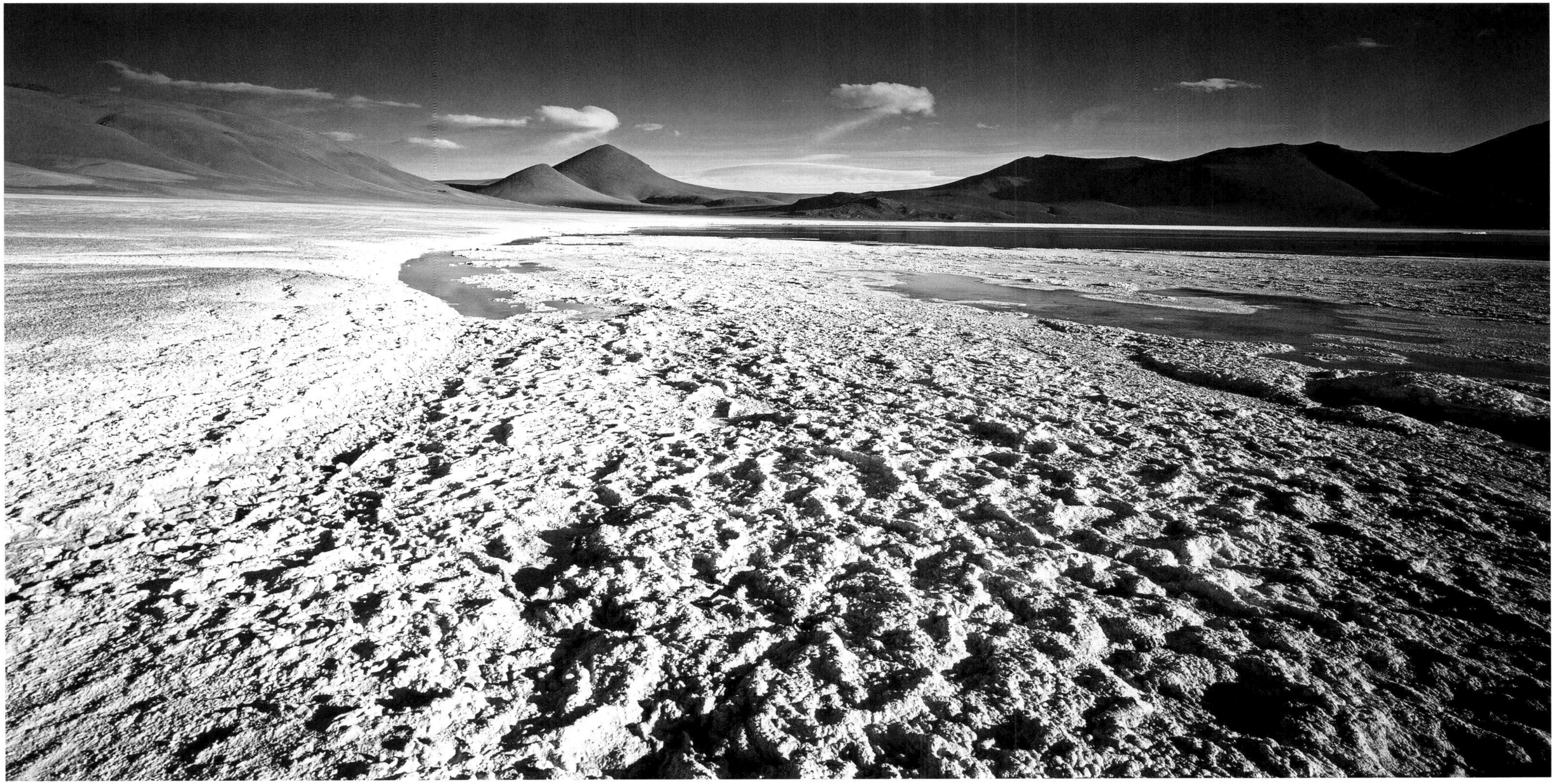

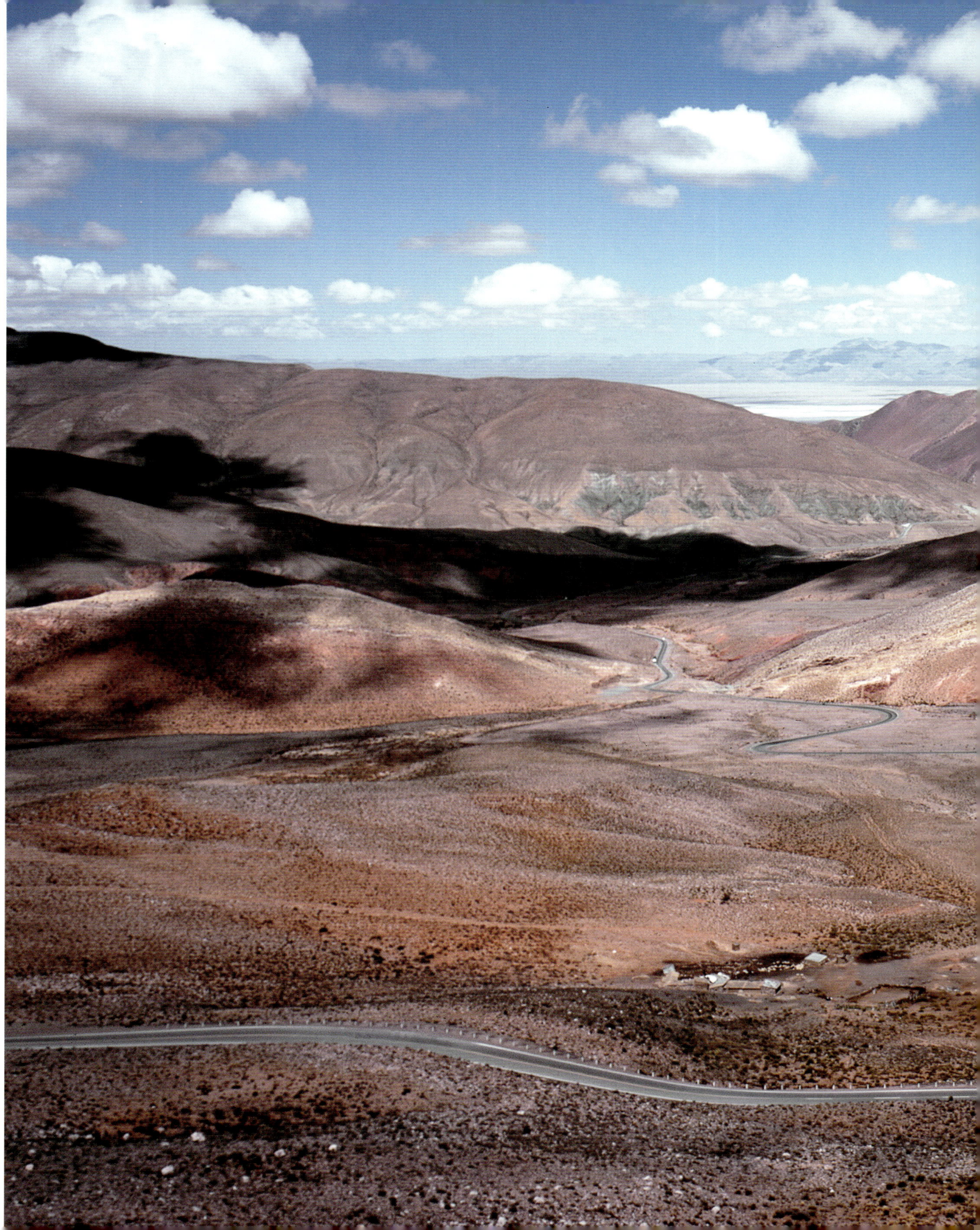

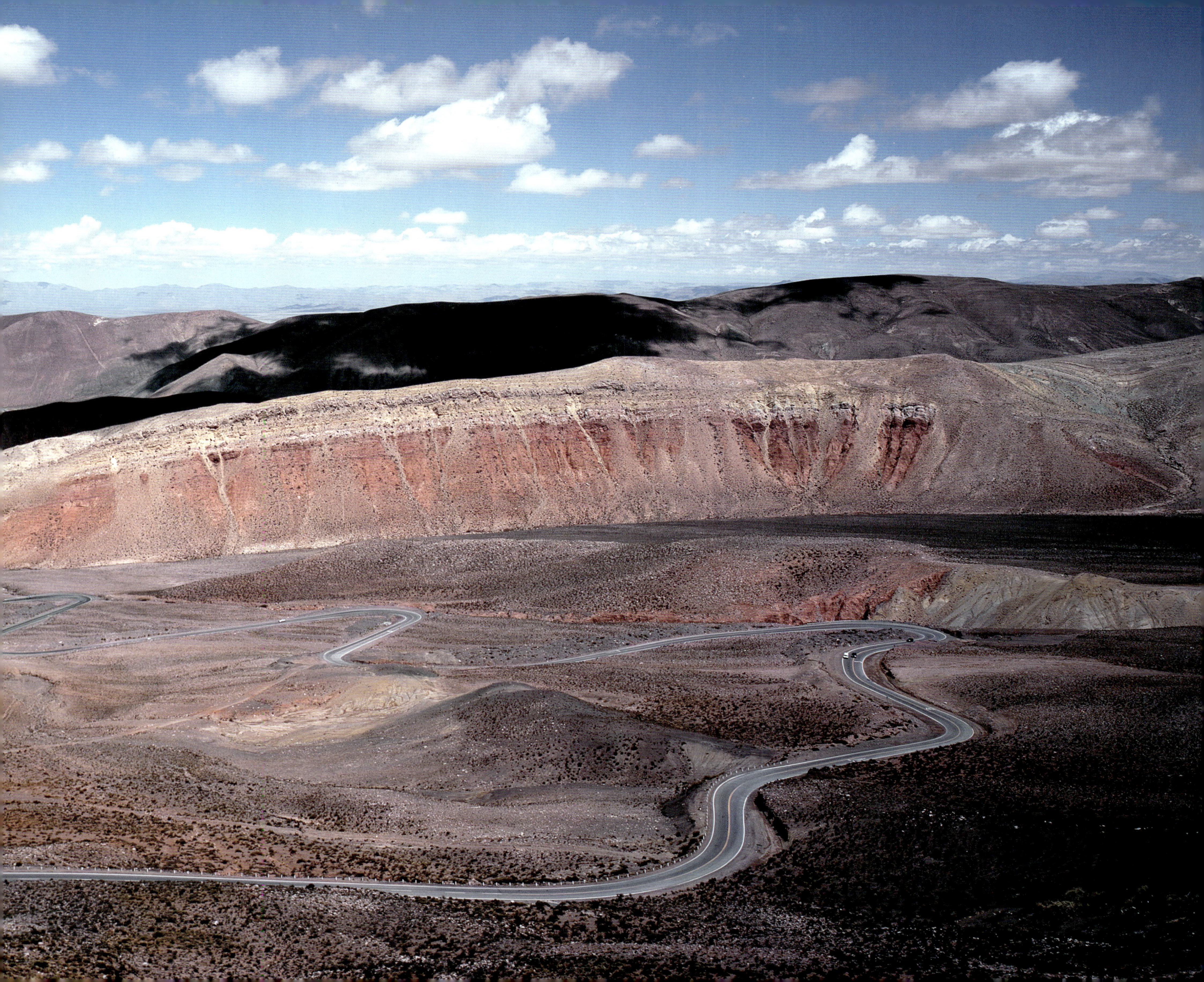

JEANNETTE PLAUT

The Thinking Eye

IN THE EARLY SIXTIES, historian Colin Rowe and the painter Robert Slutzky, described the difference between literal and phenomenal transparency.[1] For the two authors, transparency might be an inherent quality of substance or organization a distinction between literal—or real—and phenomenological—or apparent—transparency. They added that our attitude regarding literal transparency could be derived from machine aesthetics, while our feelings towards phenomenal transparency derived from cubist painting. From this phenomenal transparency, frontality, suppression of depth, contraction of space, definition of light sources, and a growing accentuation of gridding appear. Thus, transparency means the simultaneous perception of different places. By contrast for Gyorgy Kepes, another theorist of vision contemporaneous with Rowe and Slutzky, transparency implies something more than a merely optical characteristic.[2] Instead it supposes a much broader spatial order: ceasing to be perfectly clear, it becomes something that is clearly ambiguous.

Laszlo Moholy-Nagy points out that "the transparent qualities of the superimpositions often suggest transparency of context as well, revealing unnoticed structural qualities in objects."[3] In other words, through a process of distortion, recomposition and *double entendre,* linguistic transparency can be achieved. Whoever experiences one of these connections will enjoy the sensation of looking through a first plane of significance and transposition to others lying behind.

Through a catalyzing, analytical process, Erieta Attali proposes new atmospheres that generate and weave complex spatial orders and simultaneous perceptions where transparency and carefully contrasted light occasionally manage to tell a story that is conveyed continually, though not always correlatively. Elements of nature such as light, dust, rock, water, desert, and extreme landscapes, an interest stemming from her early work photographing archaeology and landscapes, seem to fuse with meticulously selected architectonic parameters abstracted by her particular eye. Her panoramic lens captures and conjugates the subtleties of a wide range of lighting qualities while careful attention to the darkness is reflected in mysterious contrasts between light and shadow. The darkness evokes a relationship with the mystery while the splendid light forms part of an exalted mixture between stone, striking shapes, spaces, and landscapes that blends into a synthetic abstraction that conveys a narrative. Erieta seems to evoke and, at the same time, provoke these narratives. Her phenomenological work poses both a contraction and expansion of space through superimposition and constant change of figure and background.

Roland Barthes once observed that the relation between a condition of sensuality in clothing is most pronounced in the edges where garment meets skin where a glove ends and the arm begins or where a sweater ends and the neck begins. Likewise, Erieta Attali's work enhances the boundaries between landscape and architecture with no intention of fusing them. Instead of focusing on predefined elements such as a door or a window, she highlights points of friction and boundaries that energize those erogenous zones. Aside from the lightness or transparency of an architectural work, it is the multiplication of these points of friction as multiple "mirrors" or reflections that ultimately invigorates the image and transforms it into living nature. In those echoes the image multiplies the points where the eye falls and creates a new situation for the observer that no longer comes from one fixed position but from various locations, situating him simultaneously from within and without.

In his well-known video "The Looking Glass" (1982), Chilean artist Juan Downey highlights a relationship between image and eye as a lengthy tension built through the comprehension of a visual phenomenon. Erieta's work undoubtedly sits along one of those edges described by Downey as a "Thinking Eye" consisting of the architectural work, removed from its objective condition as a landscape under tension.

---

1 Colin Rowe and Robert Slutzky. "Transparency: Literal and Phenomenal," *Colin Rowe and Robert Slutzky Perspecta 8*, (1963), 45–54.

2 Gyorgy Kepes. *El Lenguaje de la Visión, Gyorgy Kepes.* (Ediciones Infinito, 1969).

3 Laszlo Moholy-Nagy. *Vision in Motion,* Laszlo Moholy-Nagy (Paul Theobald & Company, 1947).

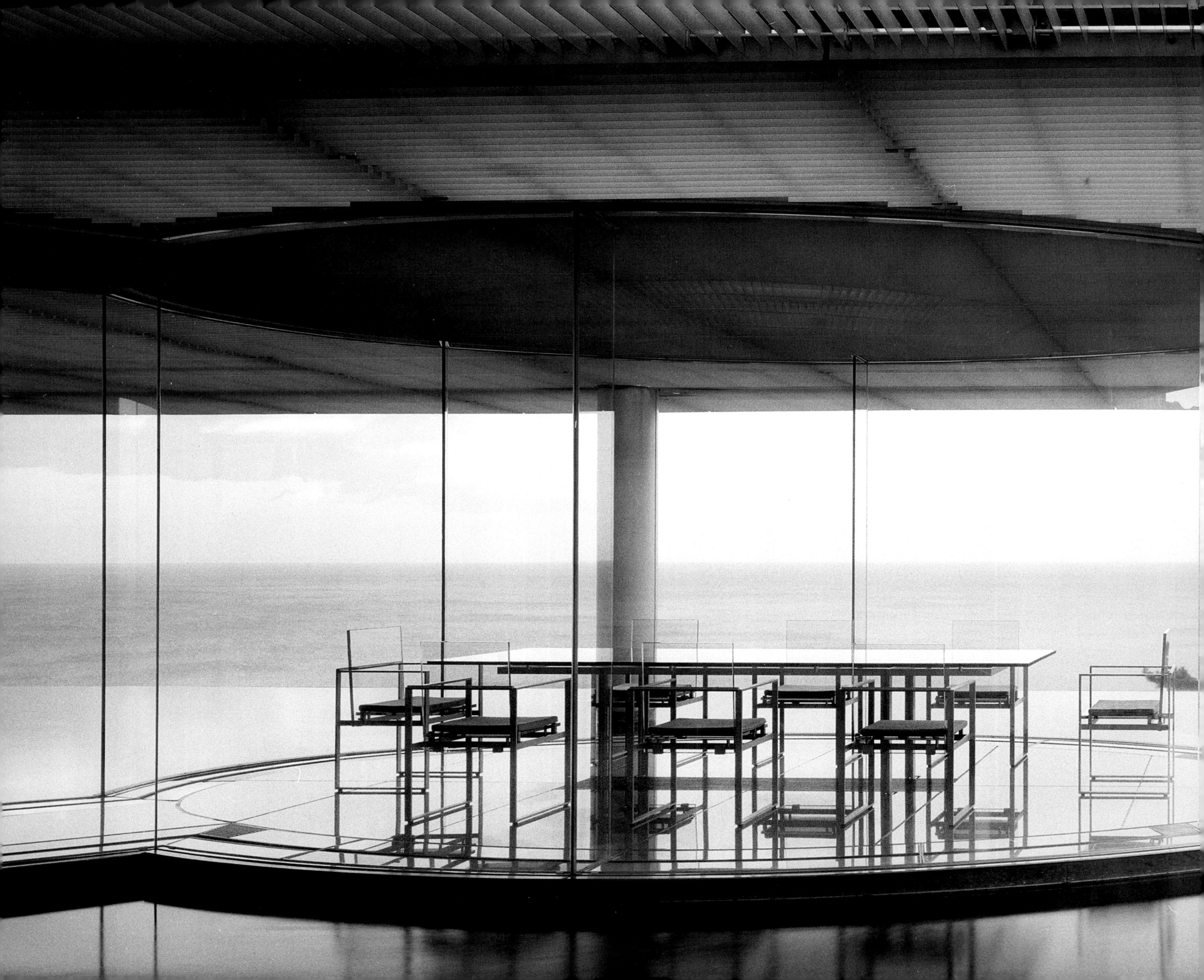

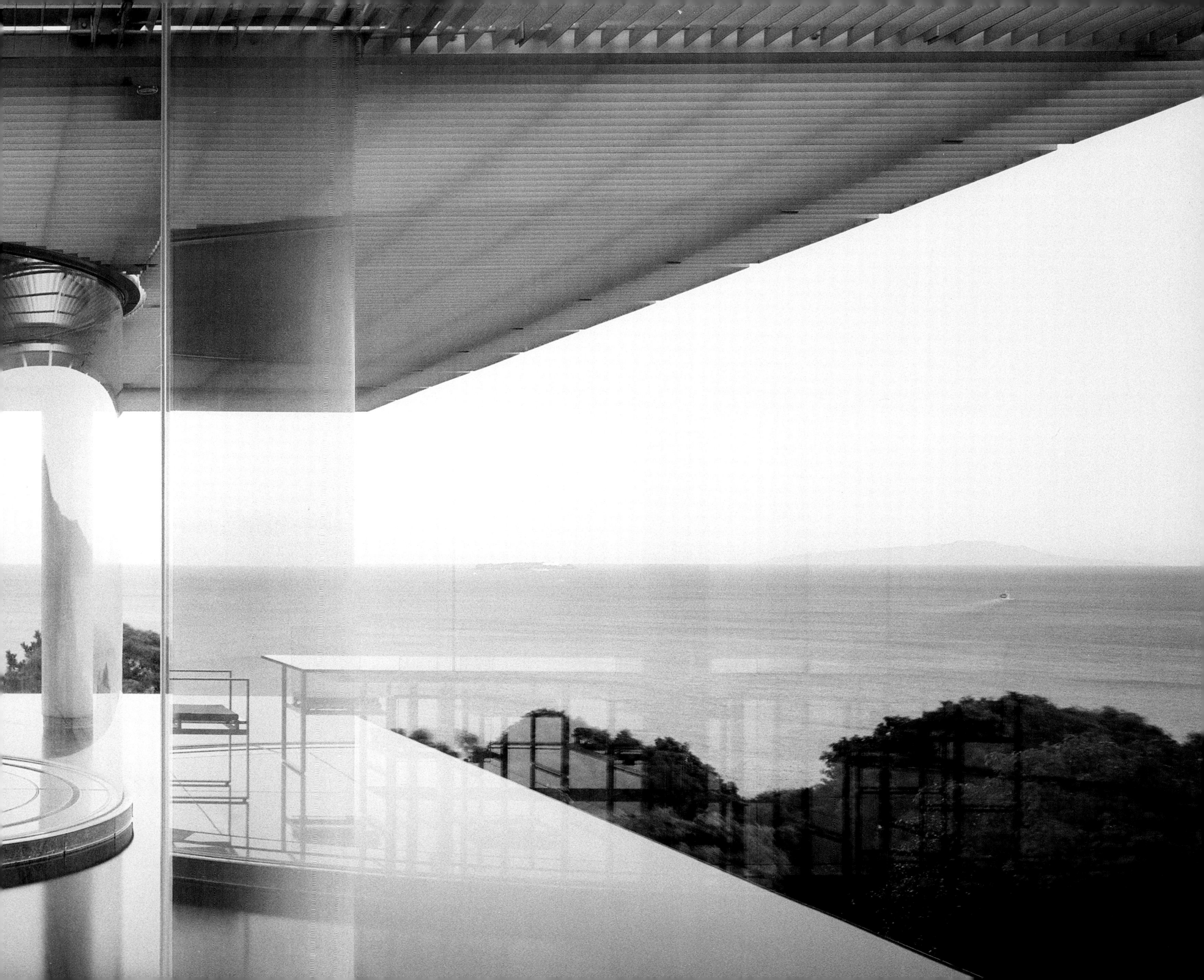

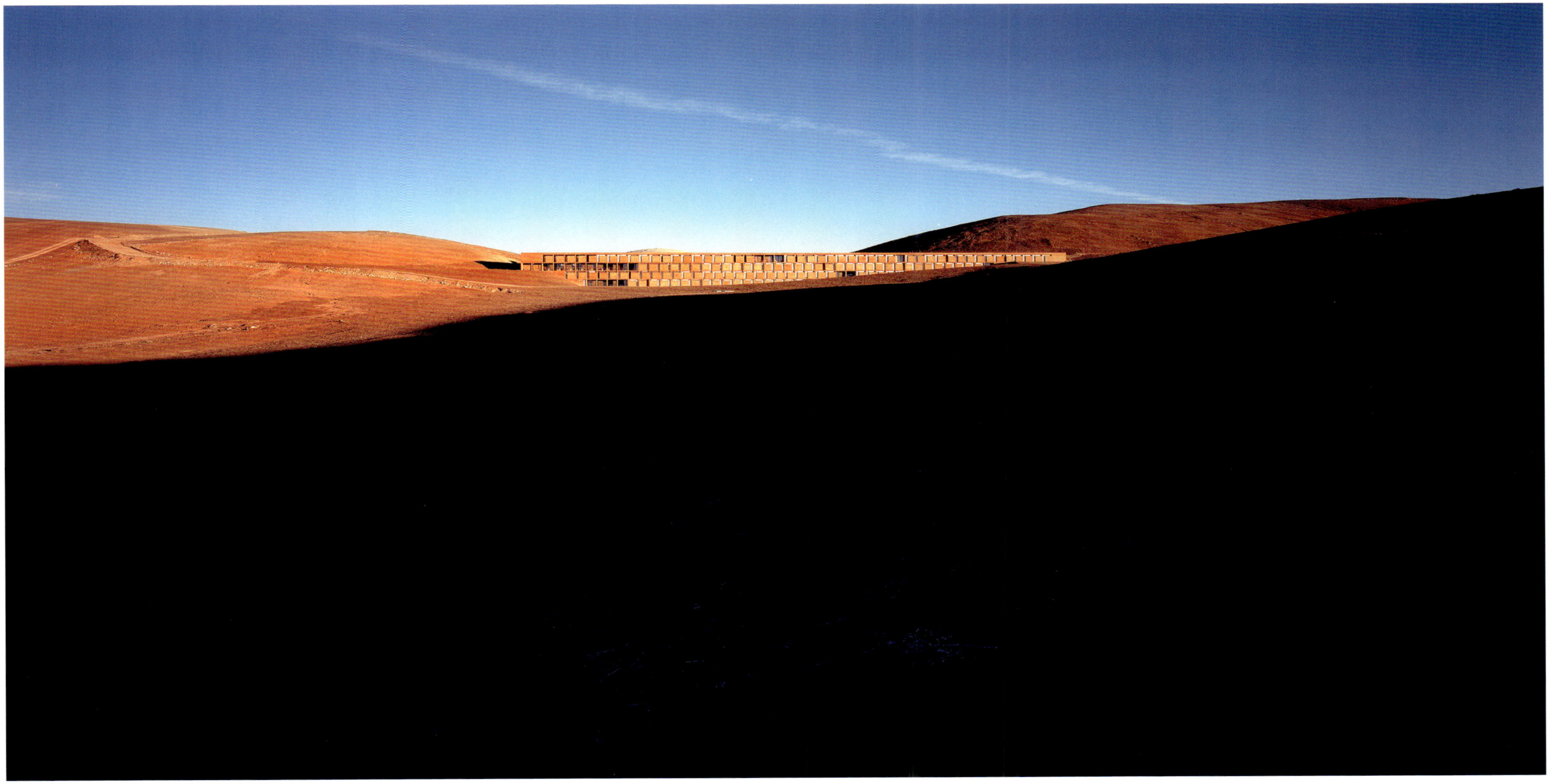

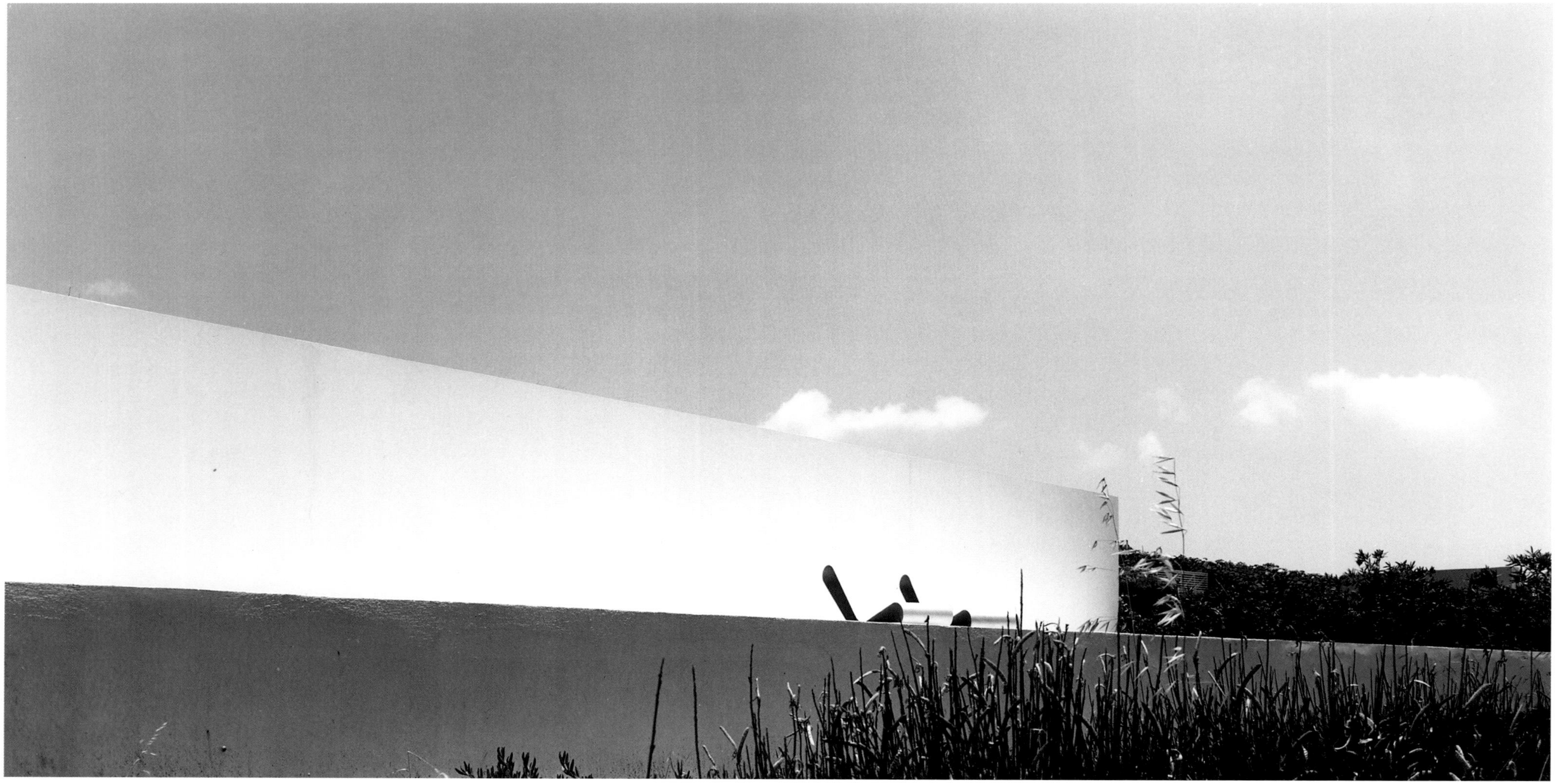

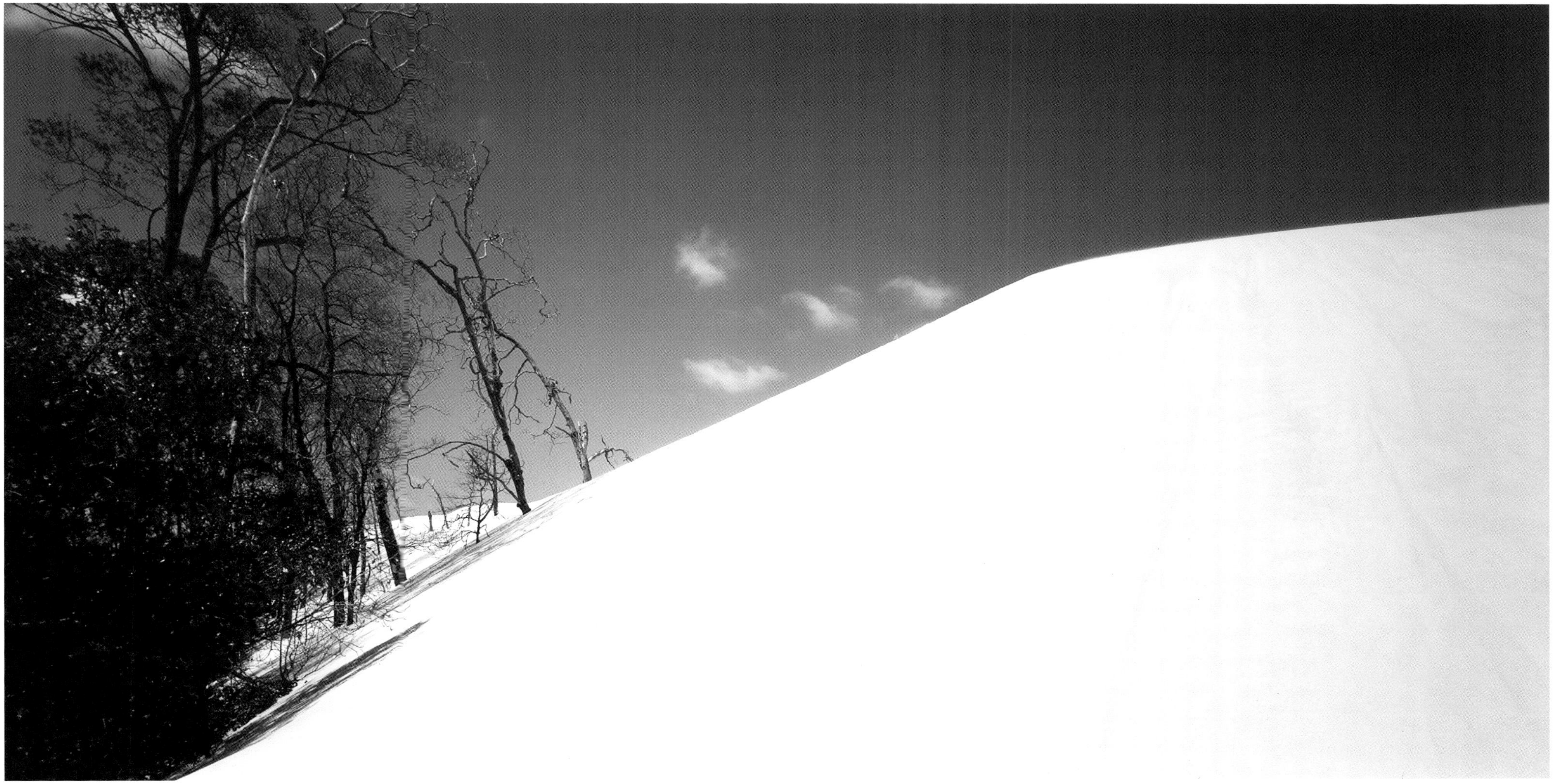

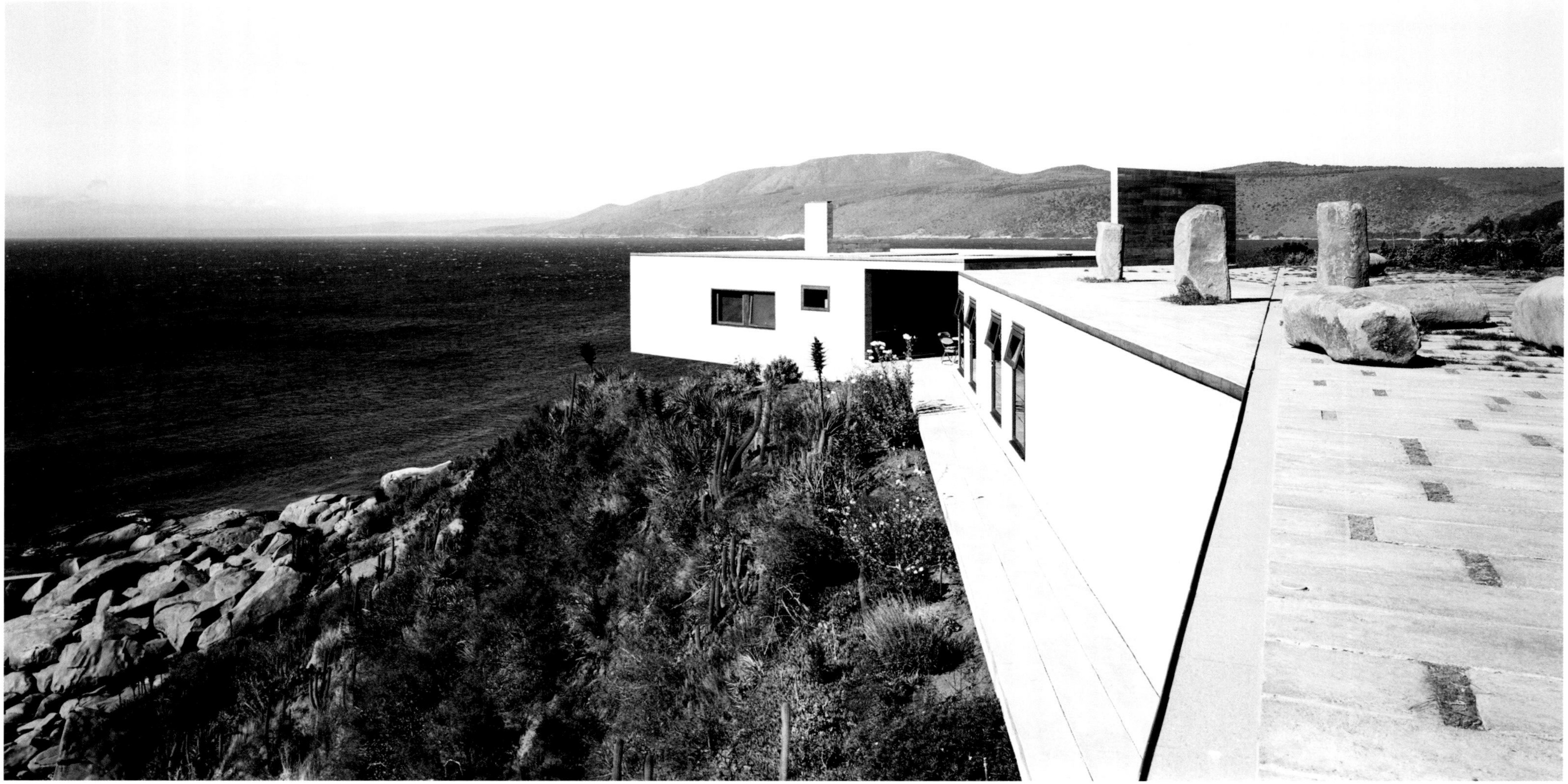

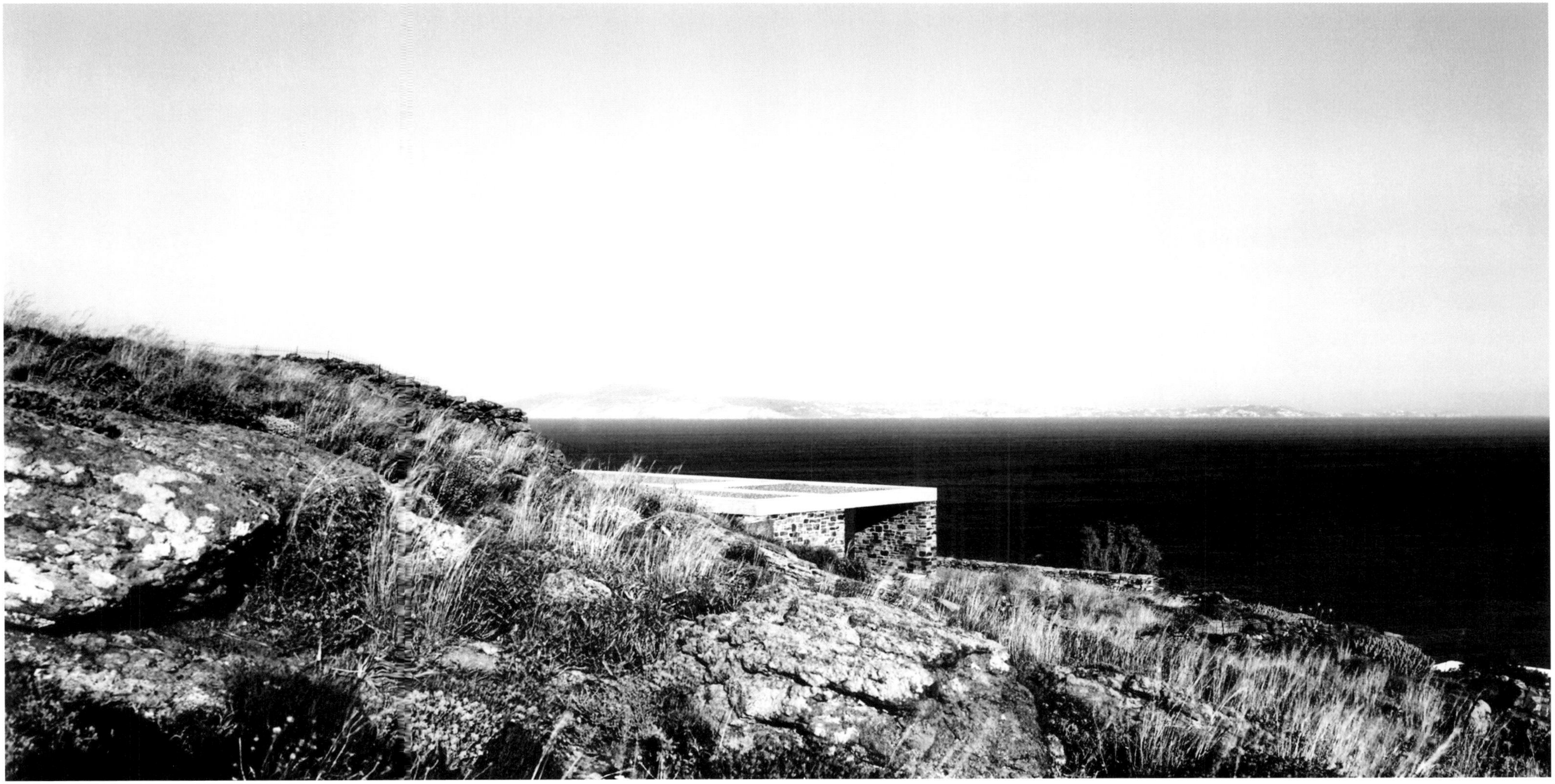

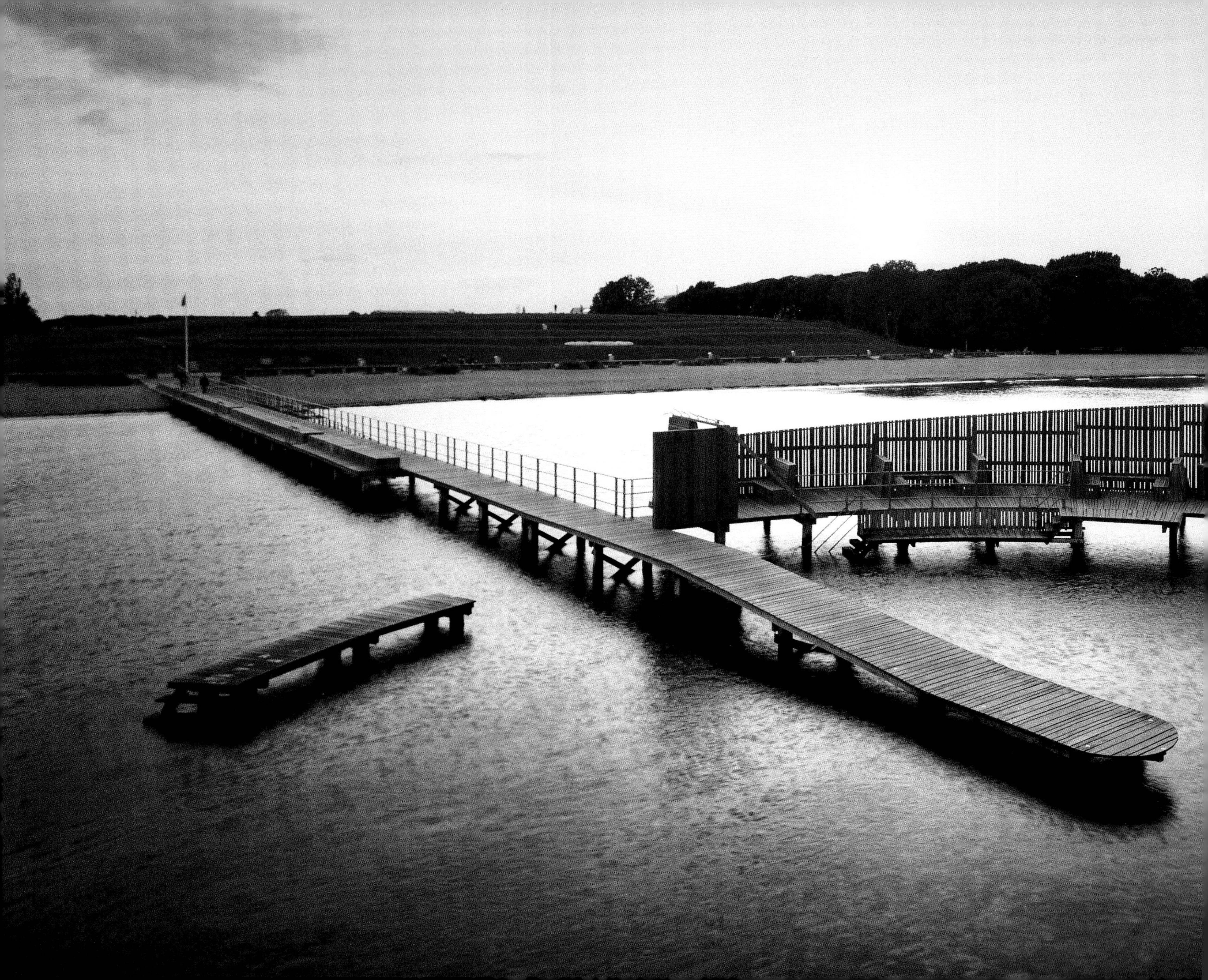

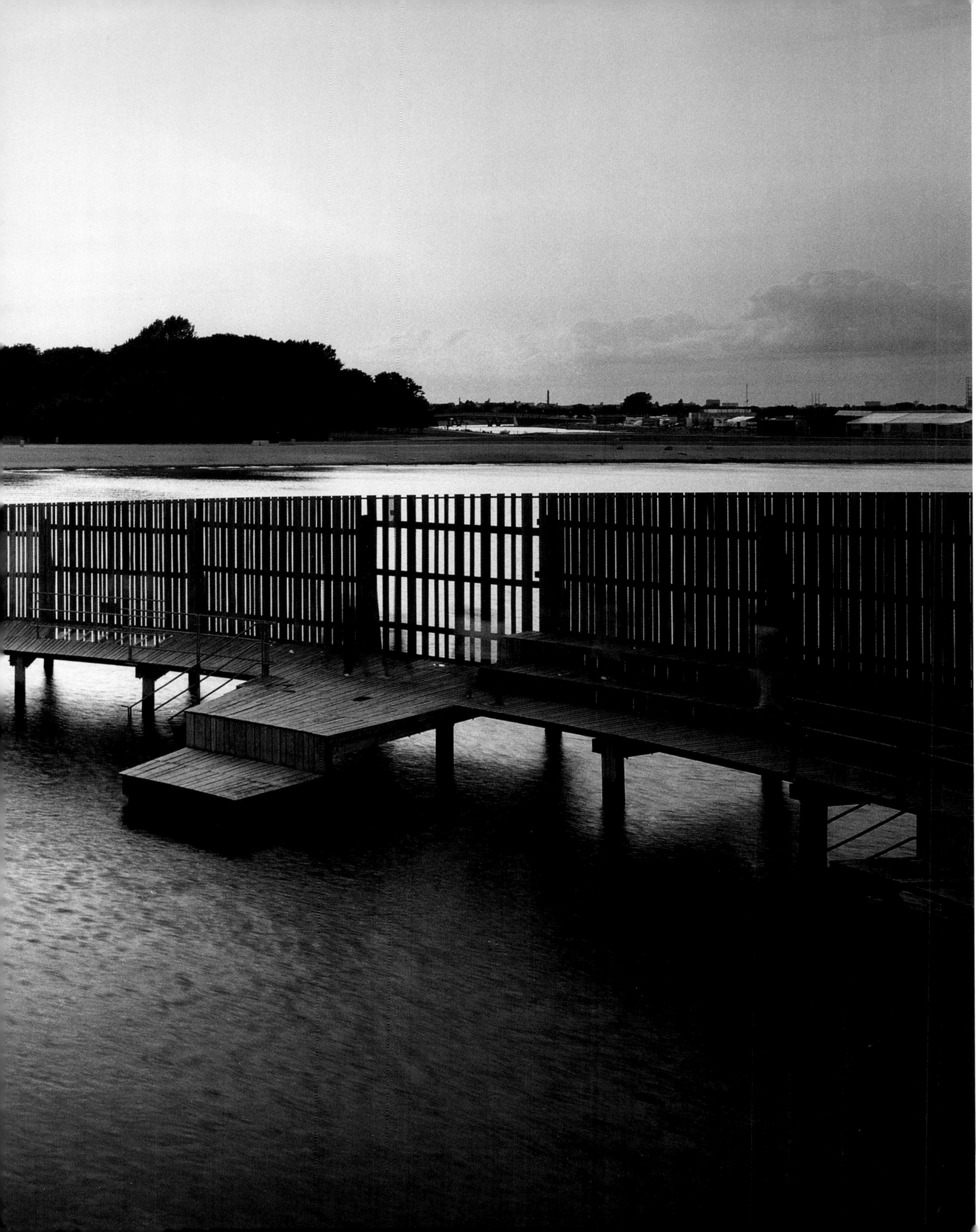

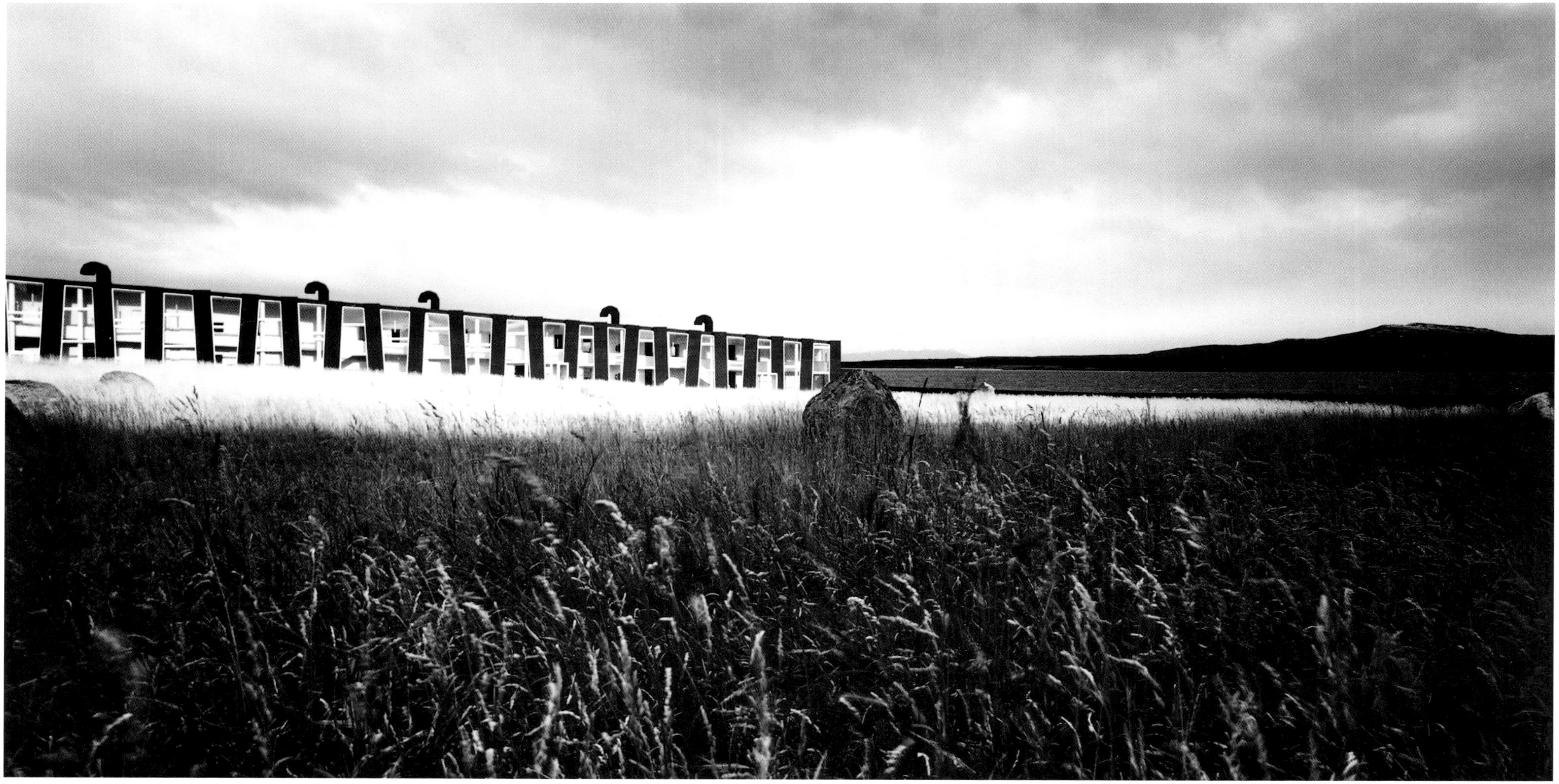

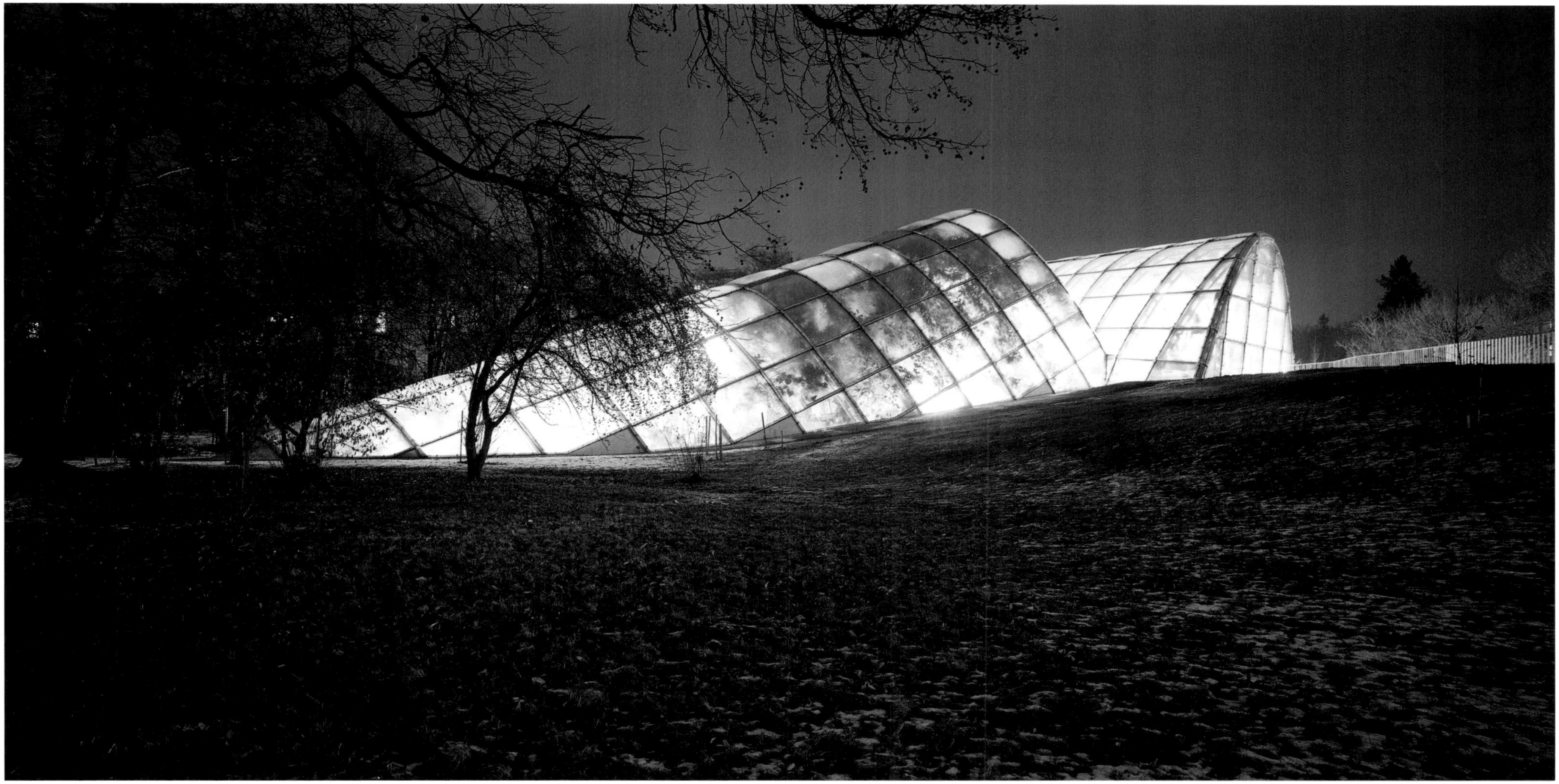

KENNETH FRAMPTON

Landscape Into Architecture

ALL OF THESE PHOTOGRAPHS, taken from a wide ranging trajectory, spanning the Americas, Greece, Switzerland, Scandinavia and Japan, are strictly devoted either to a momentary record of spectacular landscapes and seascapes invariably under the play of changing light or, alternately, to the articulation of architecture in terms of landscape and vice versa. That is to say, either the architecture is heightened against the backdrop of the landscape or, less frequently, perhaps, the landscape is sharpened through the presence of the architecture. However, with the single exception of a close-up of Angelo Bucci's Ubatuba House in Sao Paulo, Brazil, we never get a shot of architecture acting in the frame alone. In certain remarkable photographs both building and landscape seem to be engaged in a narrative interaction as though the whole image is film still. I am thinking in particular of the twilight shot of Ubatuba, where the house looks out over a stormy sky and a windswept sea while in the foreground, we can just make out the shadowy, ominous image of a woman on the roof.

Aside from the constant presence of an exceptionally sensitive, personal aesthetic in all of the panoramic images included in this anthology, there is a striking shift in mood as one passes from one rendering of the constant horizon to another. Thus a connoisseur may single out certain sublime landscape shots that have captured a special moment in a particular place where the fall of the land, the raw material underfoot, the vegetation, the sky, the sea, and the light come together to engender an unrepeatable, Heraclitan instant in the life of the earth. I have in mind such pieces as fog rising on the high ground in Jujuy, Argentina or the panorama of the multiple water falls in one single shot in the Foz do Iguaza, Brazil, or Coney Island seen in twilight across a ruined breakwater, or the photographer's beloved moonscape of the Atacama Desert in Chile.

In many of the finest images included here architecture is heightened through landform, light, and material, from Takis Zenetos's Lycabettus Theatre in Athens to Auer + Weber + Associates's Eso Paranal Hotel in the Atacama Desert, and from Aurelio Galfetti's own house in Paros to German del Sol's Hotel Renovation in Patagonia. Attali's sweeping image of a house in Santa Monica, Uruguay designed by Mathias Klotz featuring turbulent, wind-swept brush in the foreground is surely one example where landscape and architecture break even with one sharpened by the other and vice versa. In other pieces the landscape overwhelms the built form and reduces architecture to the trace of a tectonic episode that could easily be dispensed, so that in the end the intent behind the juxtaposition borders on the obscure. On the other hand in many of Attali's most recent photos an exceptional level of metaphysical abstraction is attained, as in the foregrounded mass-form of a concrete vacation house built to the designs of Nikos Ktenas on the island of Samos. In a similar manner, certain interiors take on an uncannily abstract spatial character as in Marcos Acayaba's House in Jardin, Sao Paulo or the equally abstract interior of Klotz's Techos House in Villa la Angostura in Argentina. Then there are exceptional images in which dense architectonic cubic forms set off of massive land forms before the sea as in Attali's exceptionally dramatic image of Pezo von Ellrichshausen's cubic Poli House on the Peninsula de Coliumo in Chile or in her medium ground shot of Agnes Couvelas's more vernacularly inflected house-form with battered, cement rendered walls on the island of Santorini. Surely the most magnificent shot of the entire collection is Attali's take on the so-called Lookout Point in Stegastein in Norway. Although for the absolute framing of a seascape through the agency of architecture it would be hard to surpass the classic power and calm of her vision of Kengo Kuma's Water/Glass House, set before the expanse of Atami Bay in Japan. Lastly, there are three images in this collection that somehow speak of a going to ground on the part of the photographer. The first is an uncanny, primordial shot of an overrun cemetery in Kamakura, Japan while the other two are different takes on the telluric sweep of Dimitri Pikionis's earthwork laid over the Philopappou hill in Athens which, although by no means the last shot in this book, brings this current collection of Attali's photographs to a fitting, momentary close.

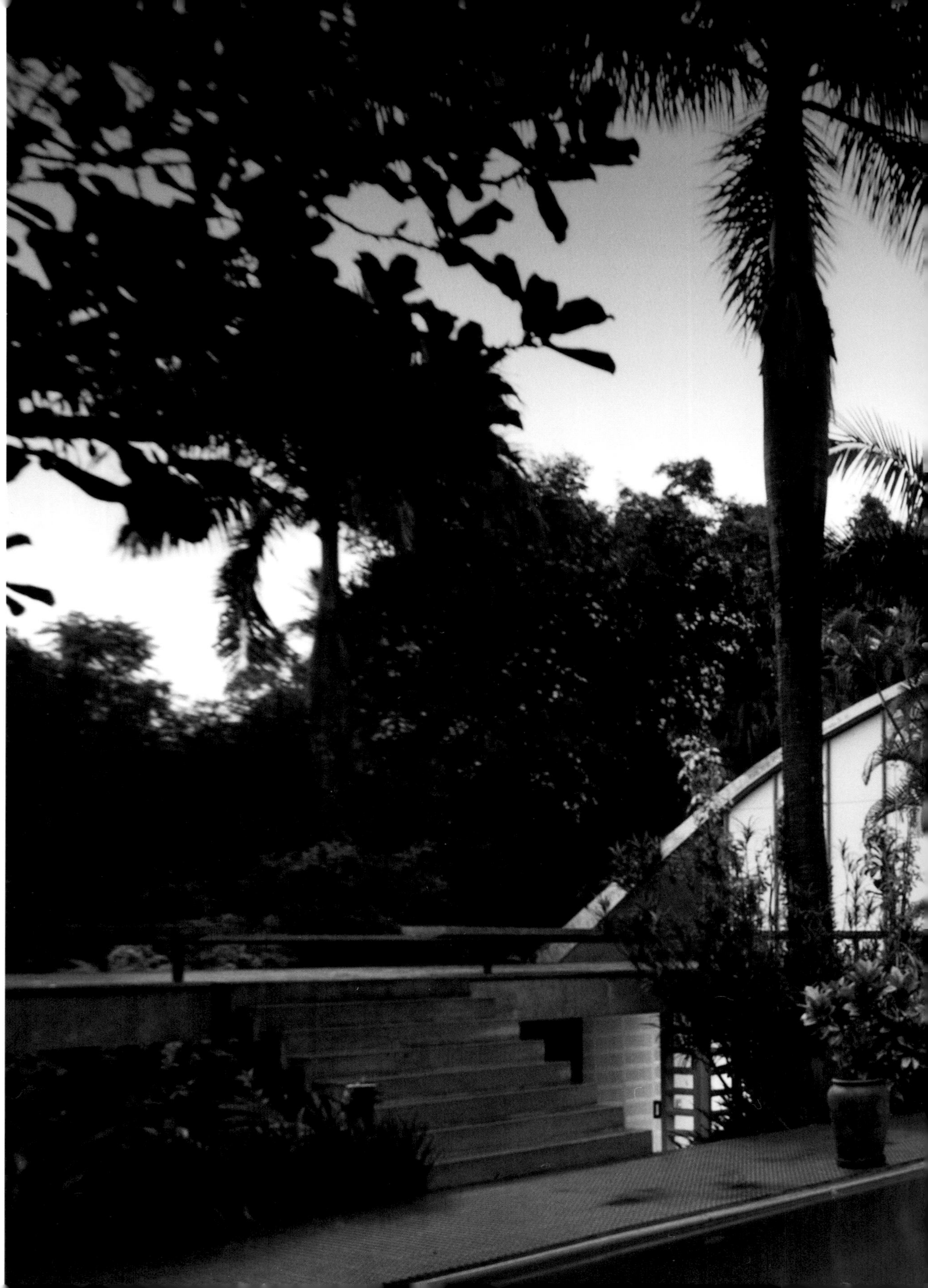

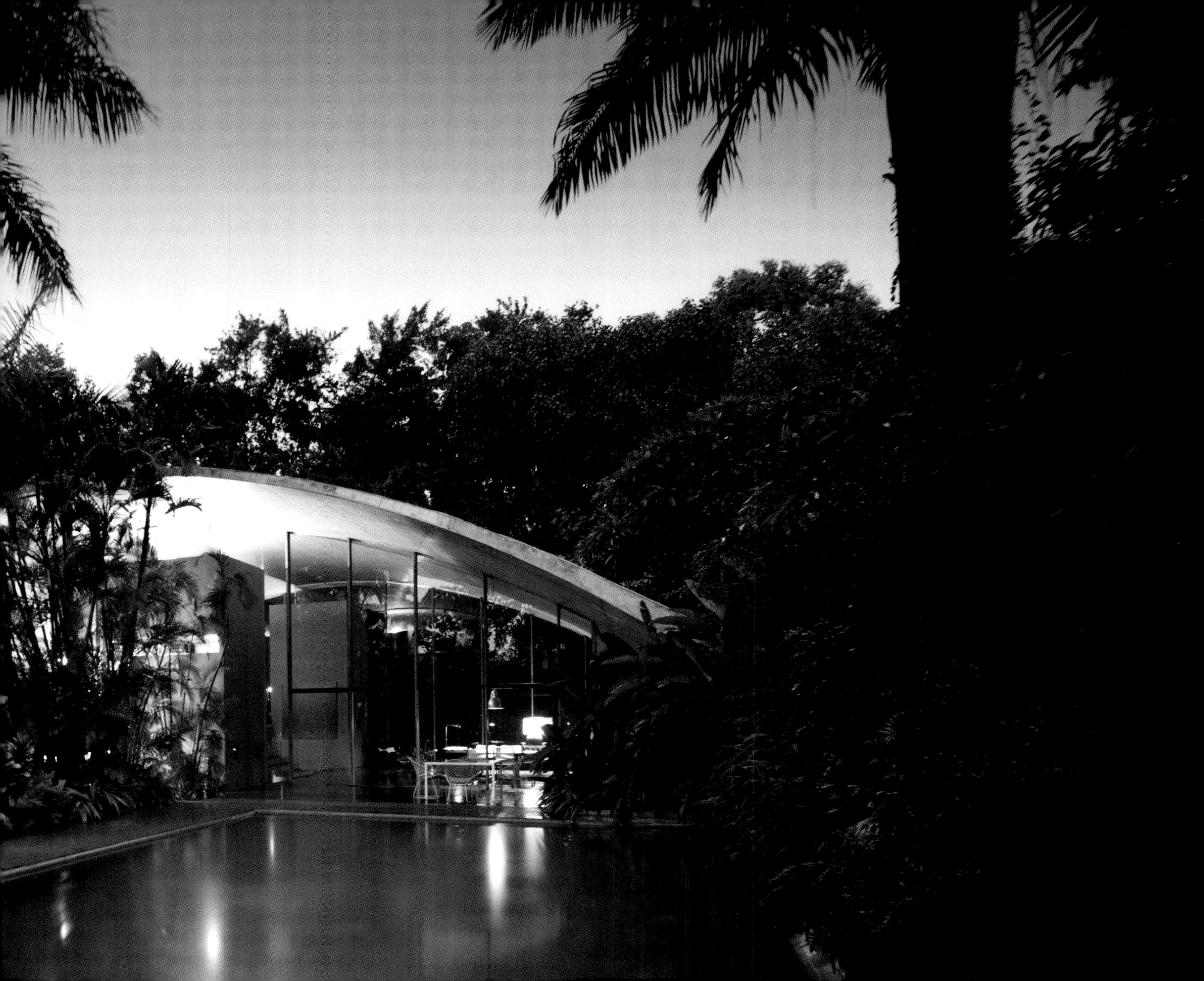

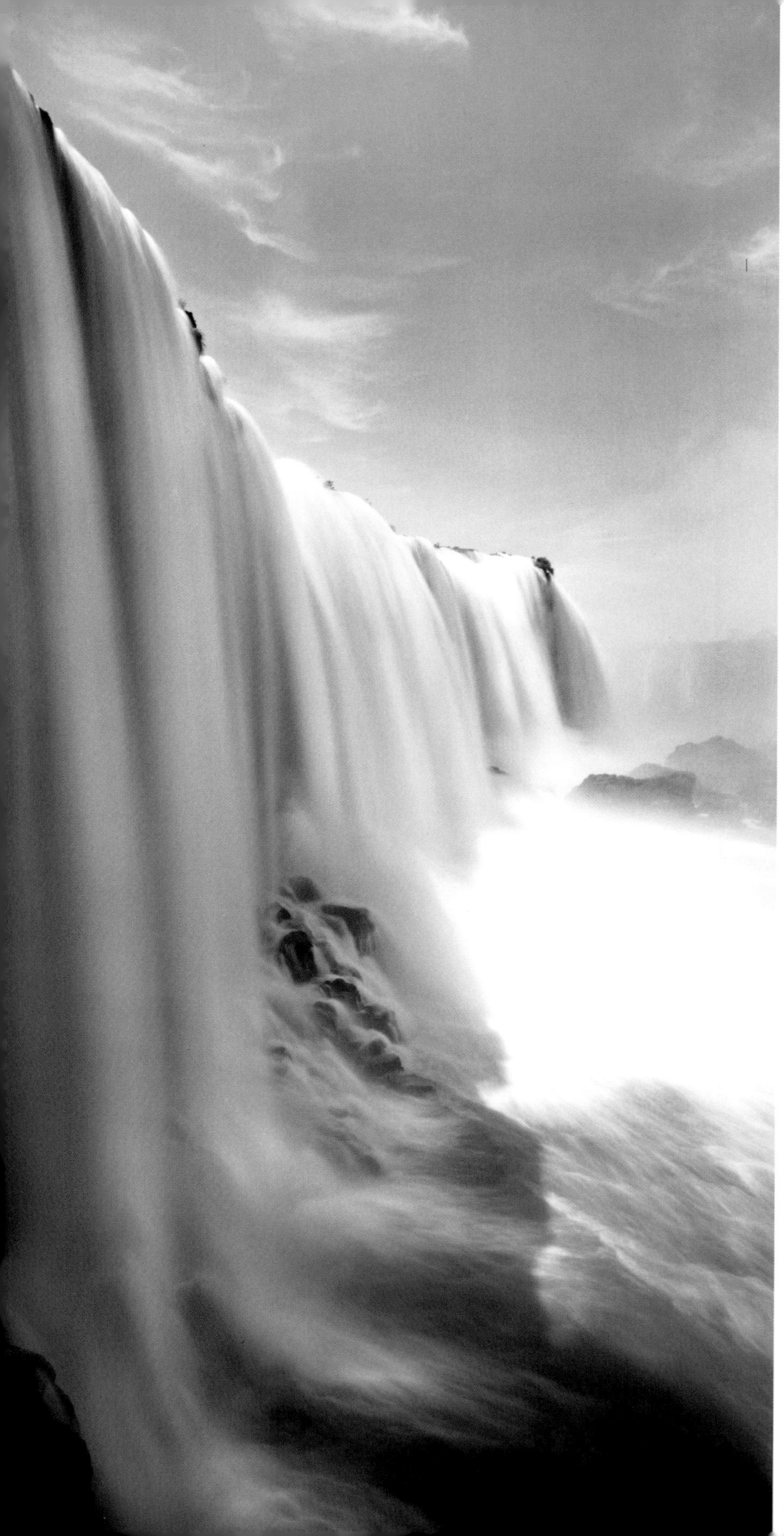

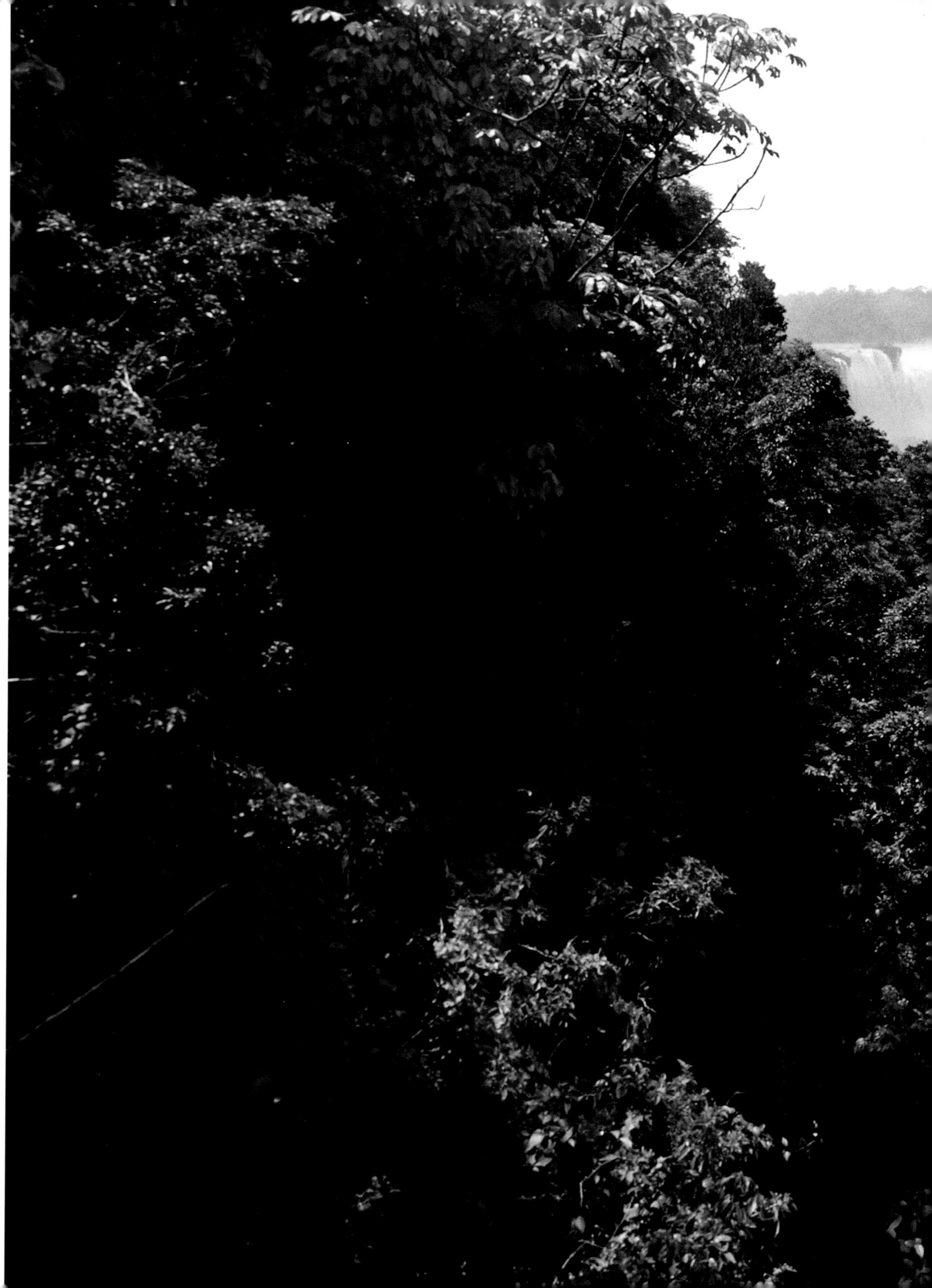

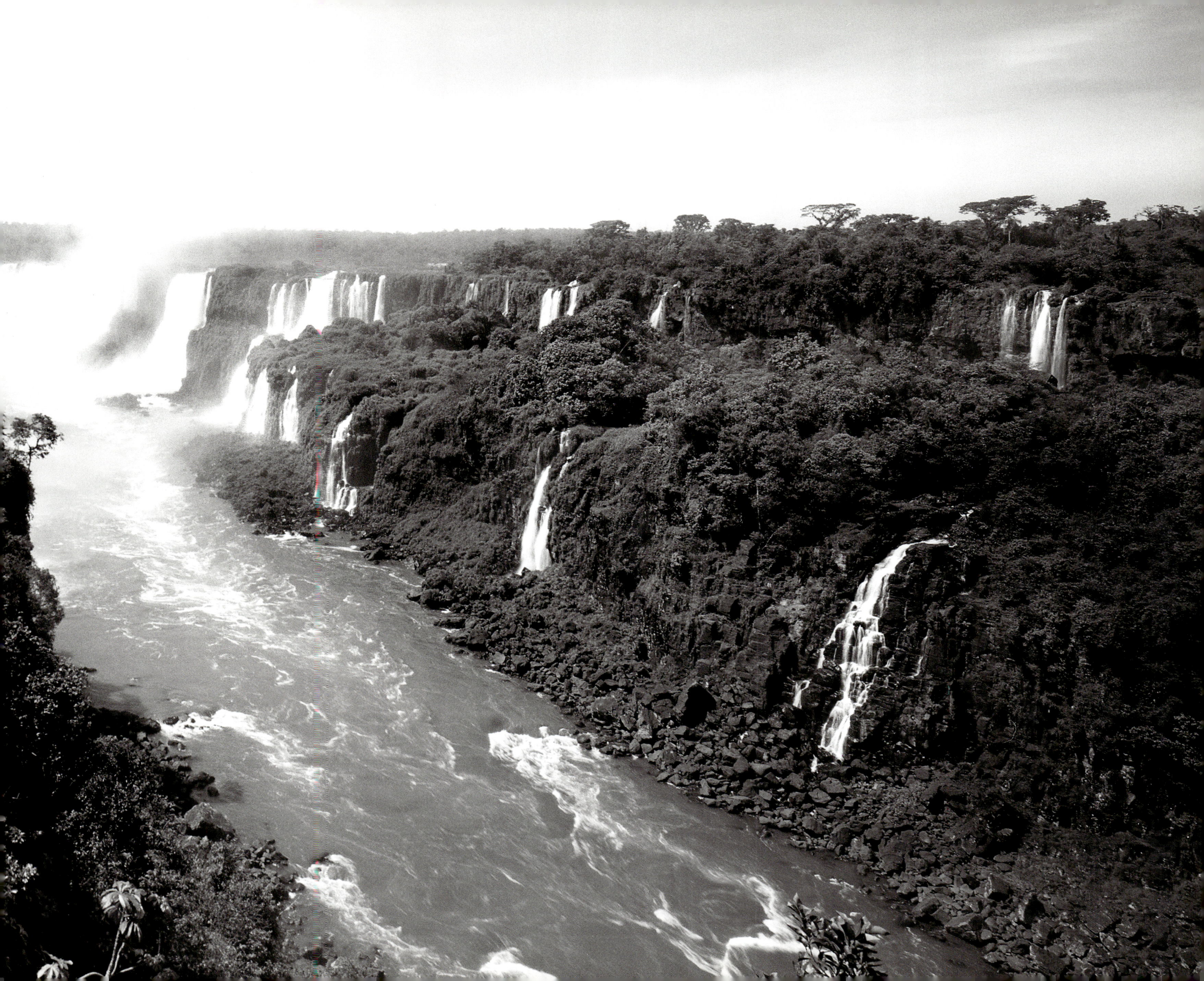

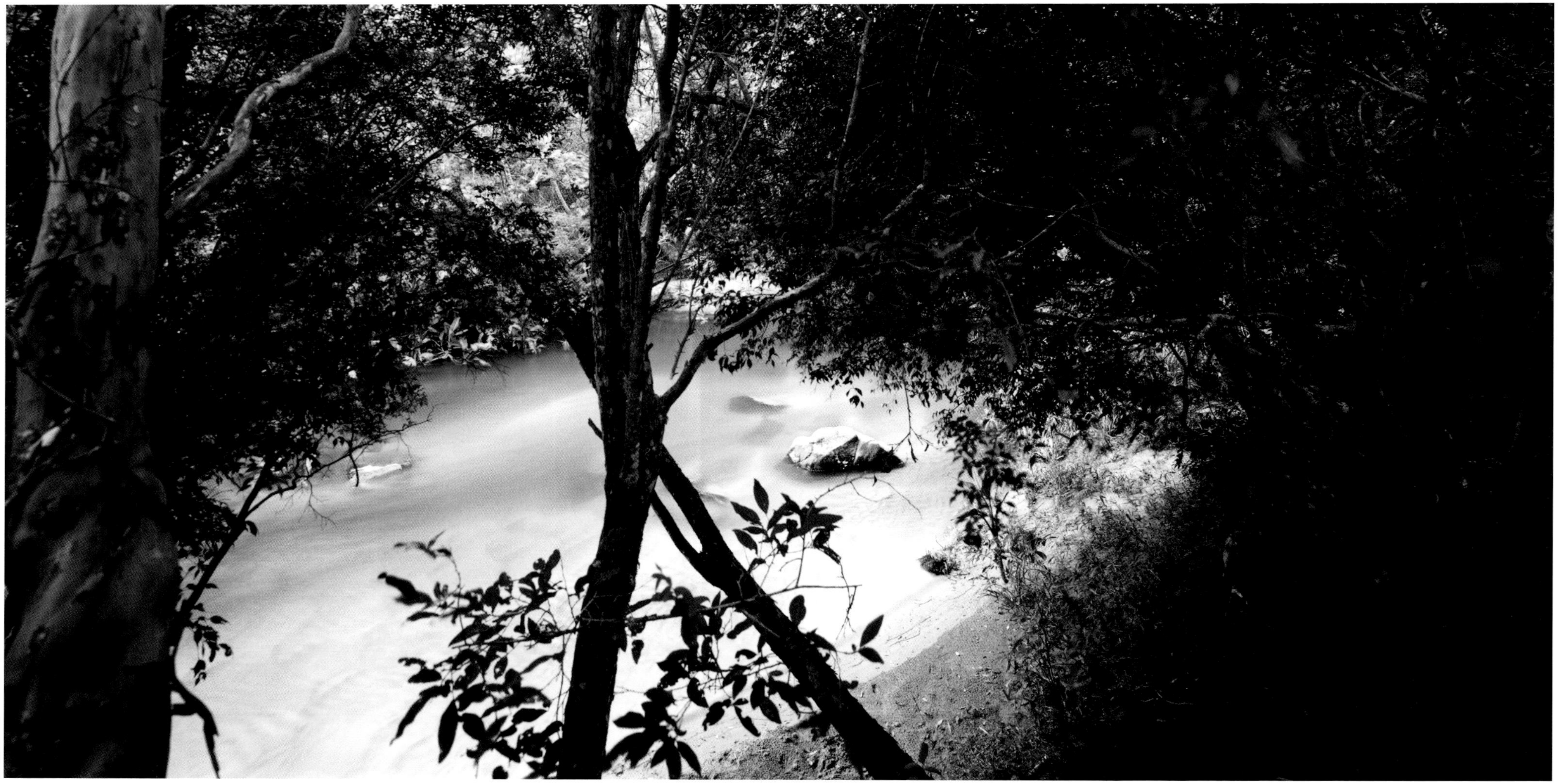

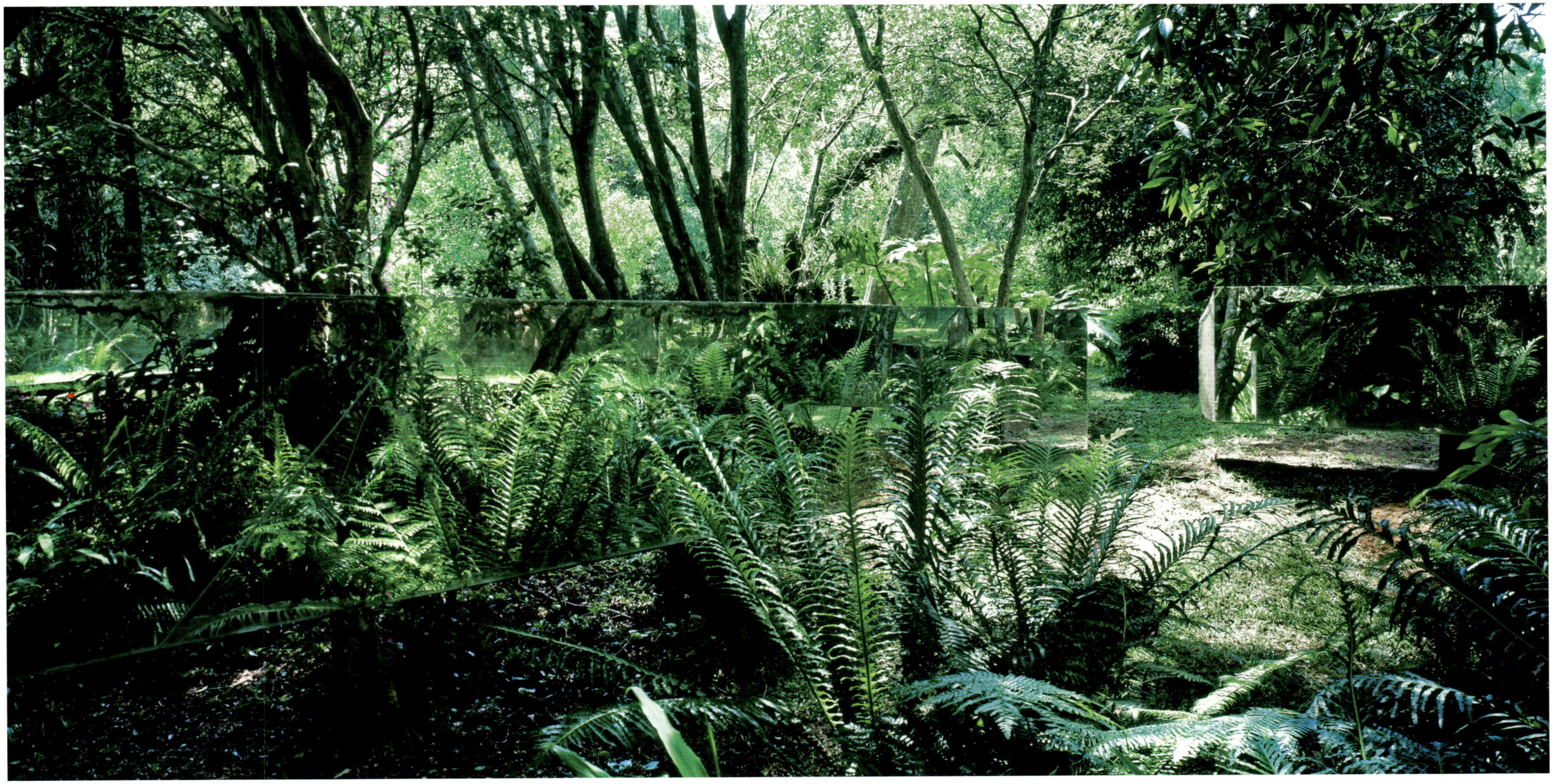

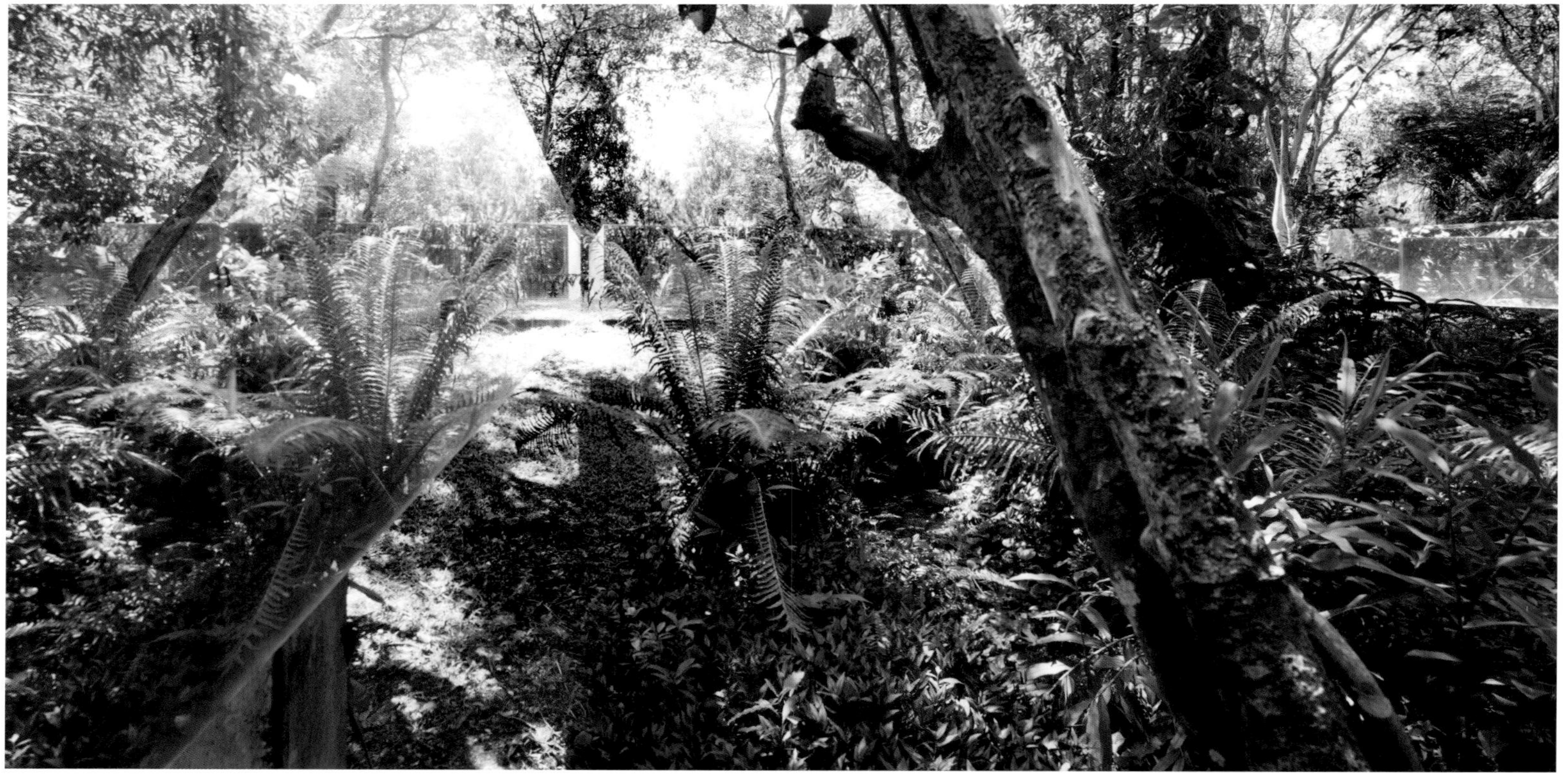

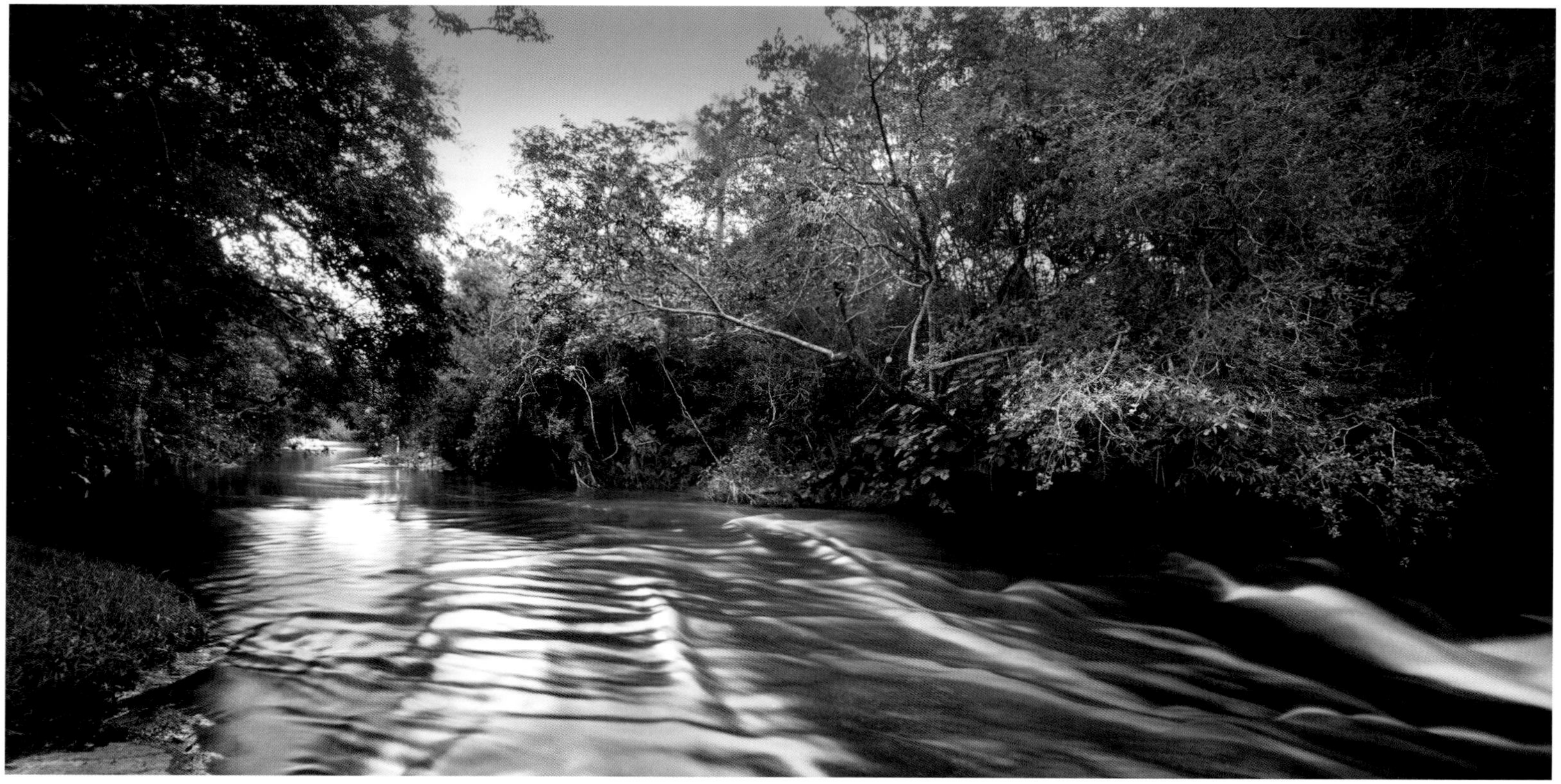

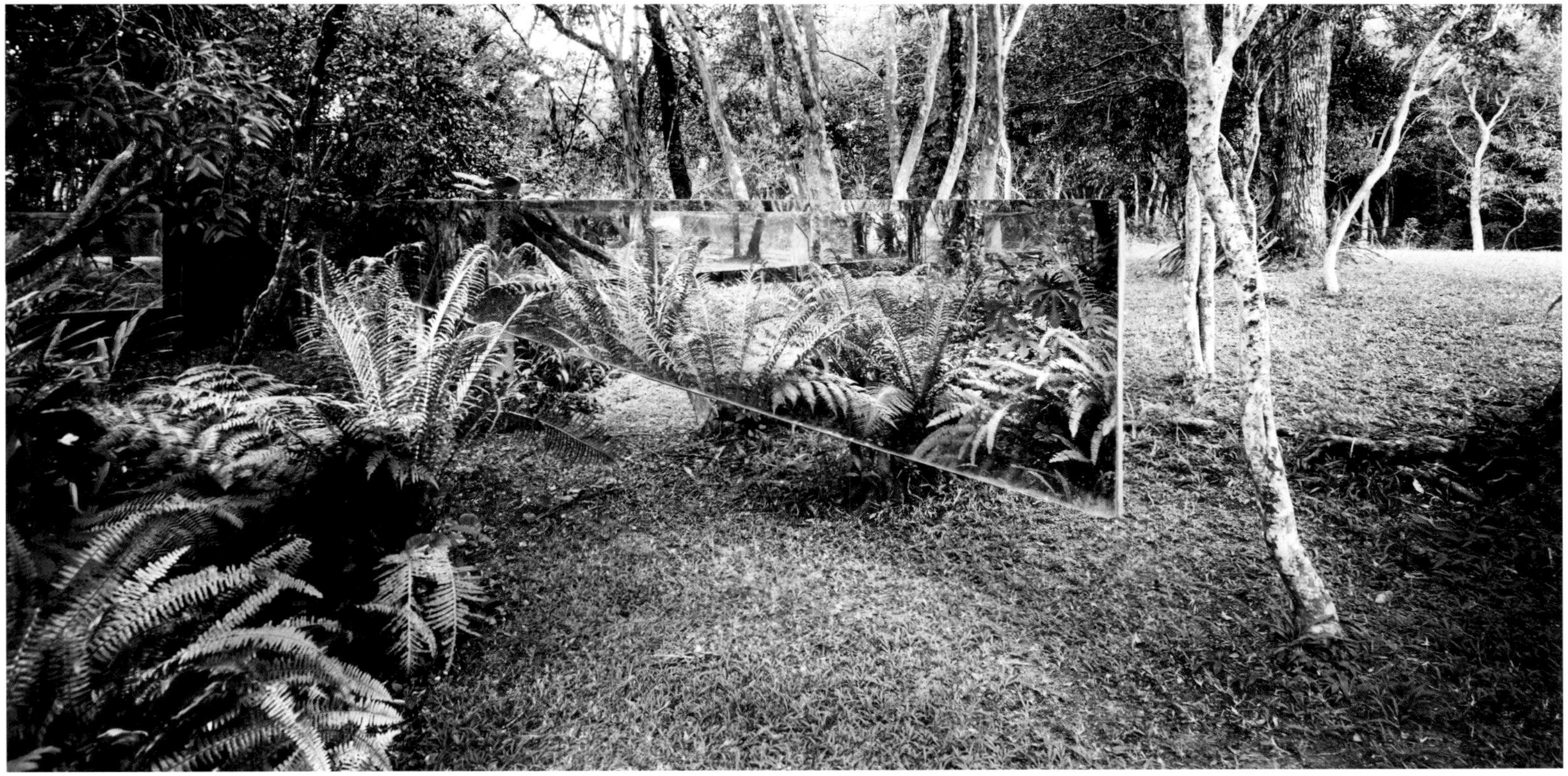

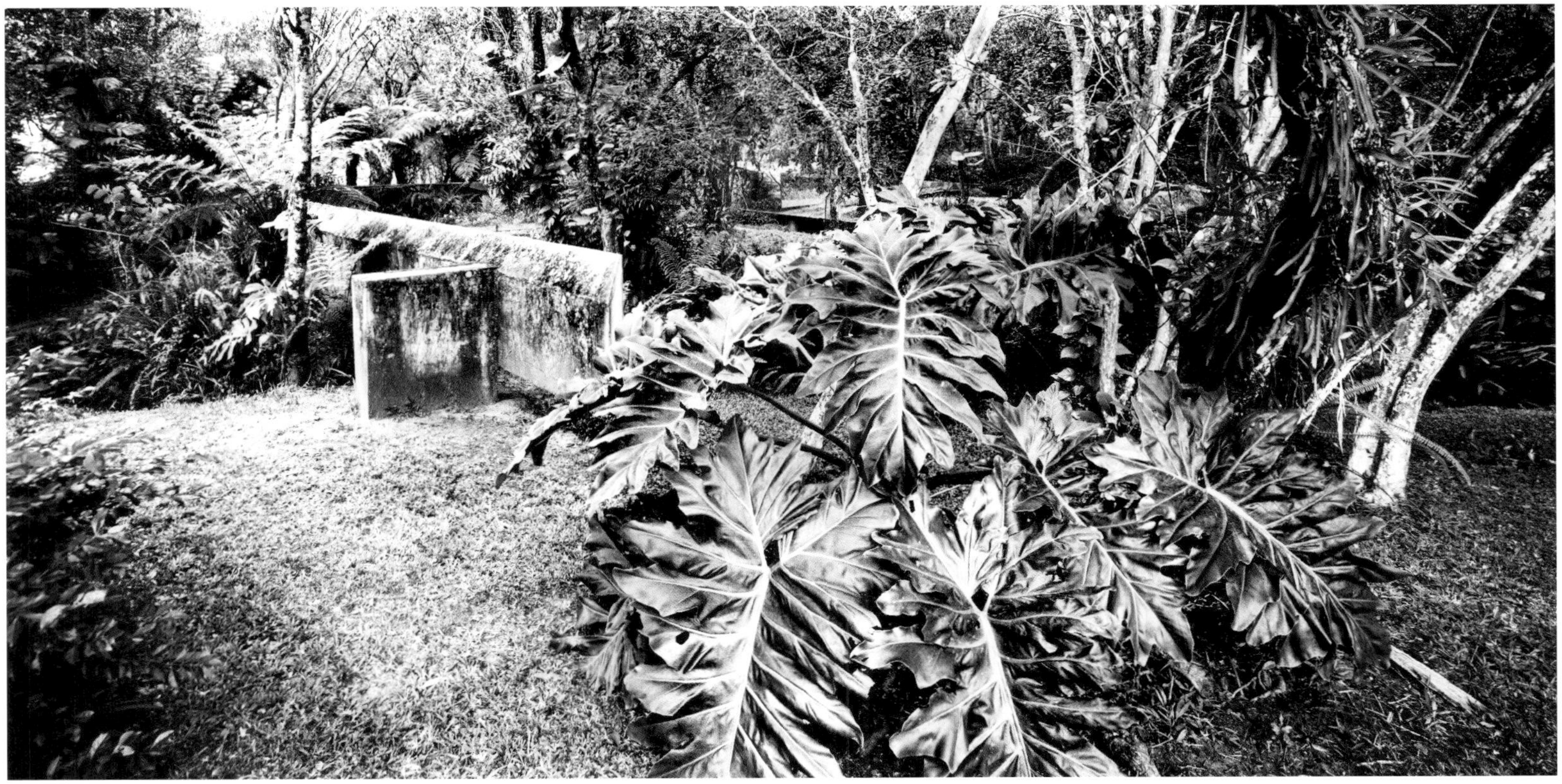

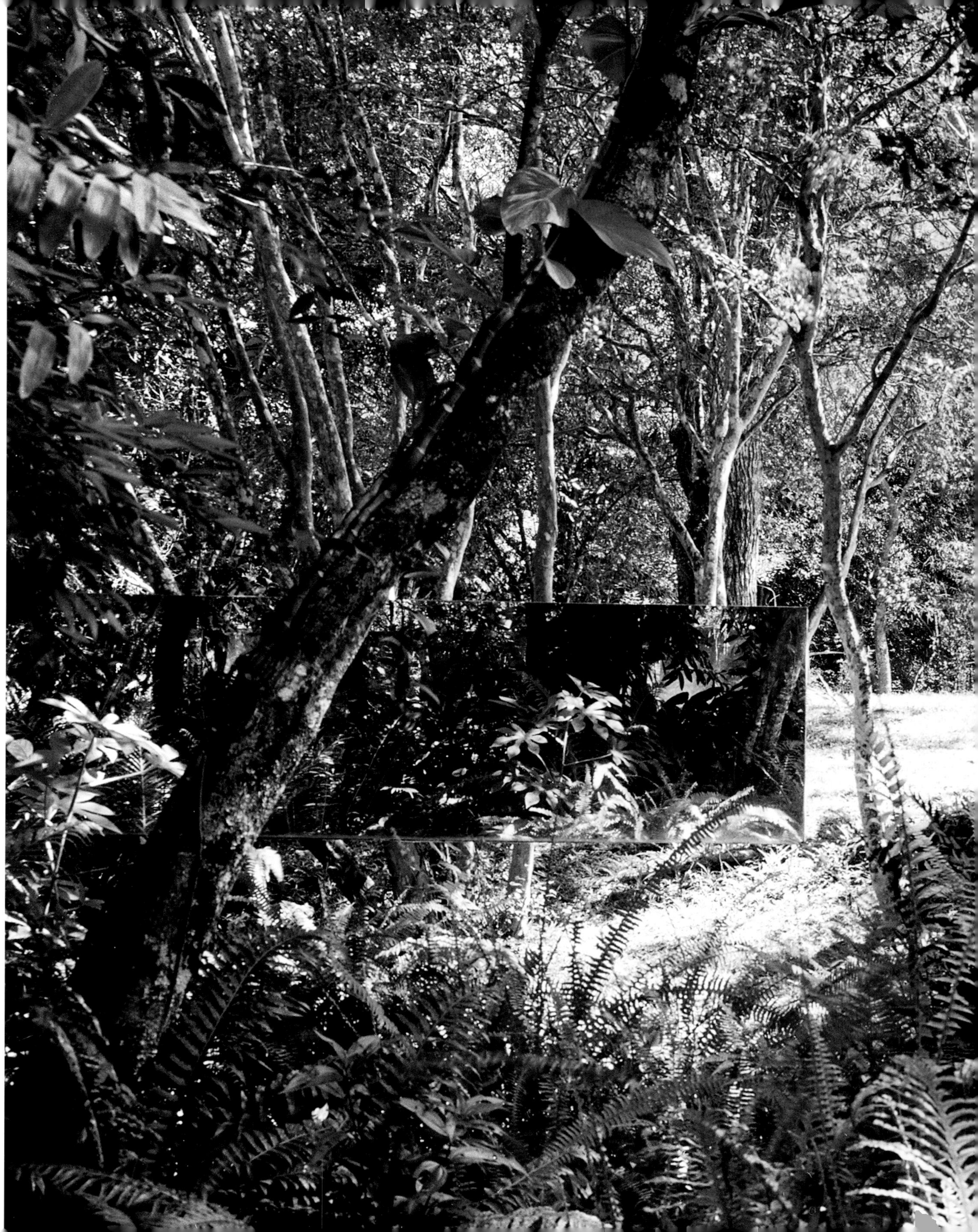

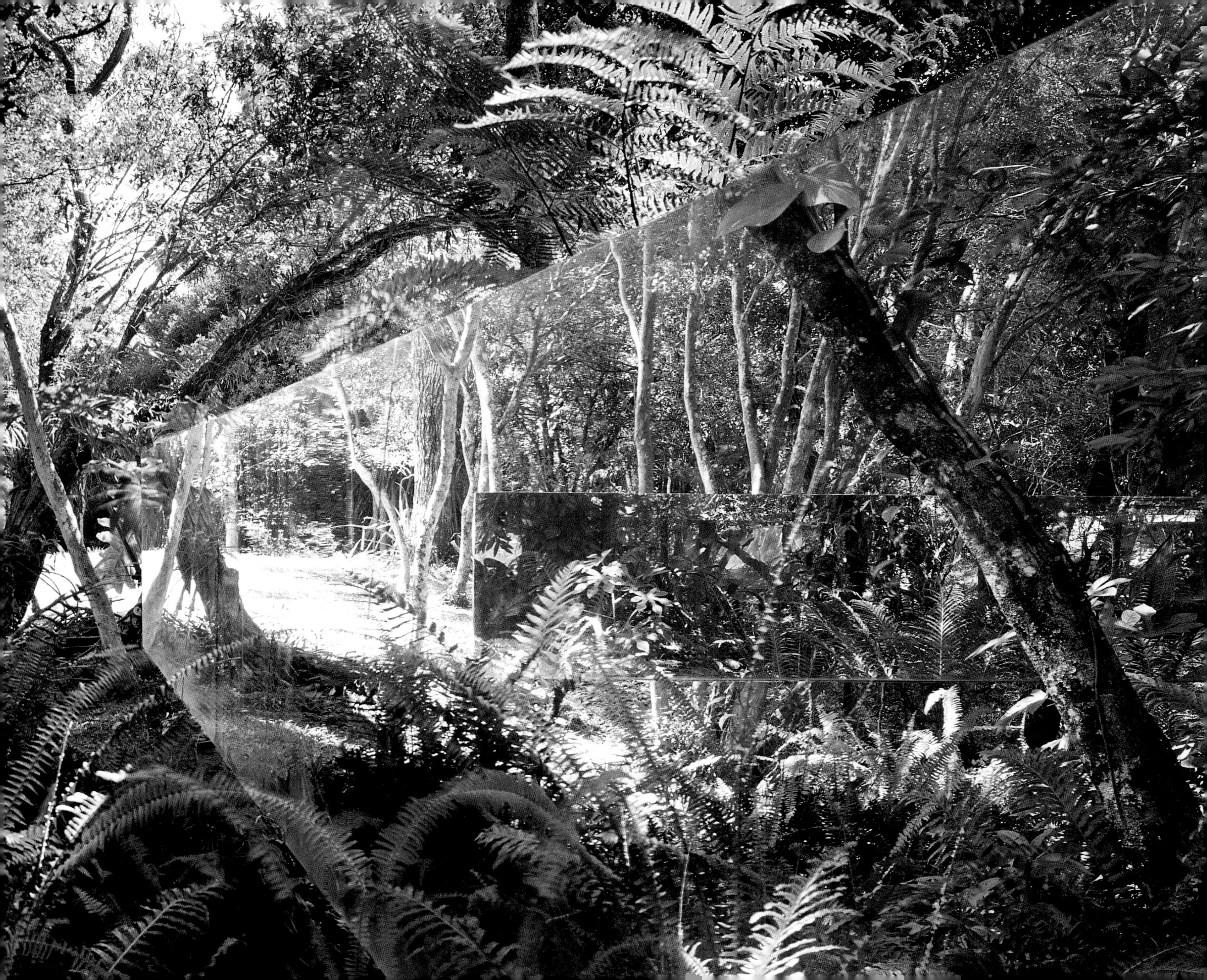

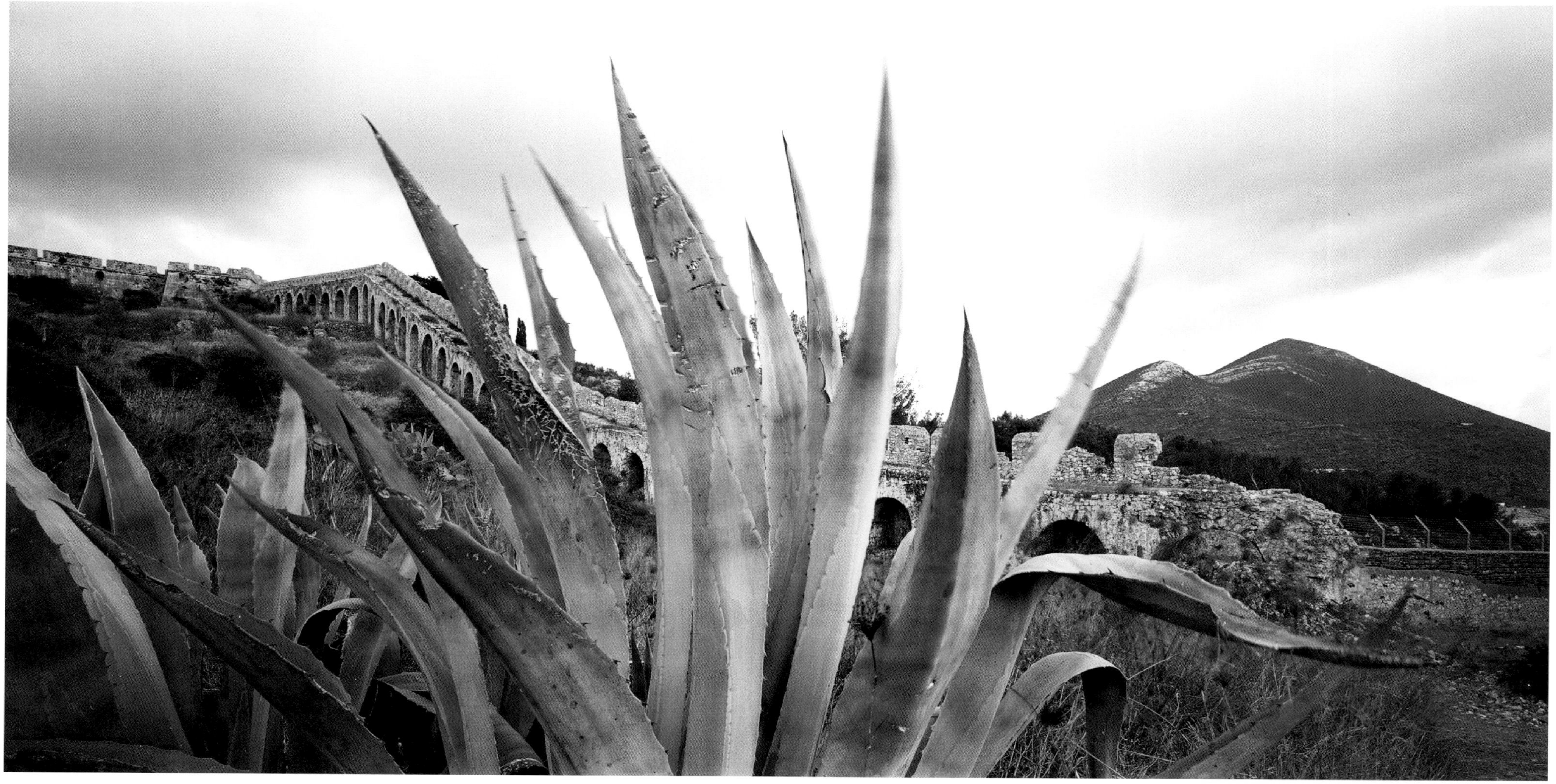

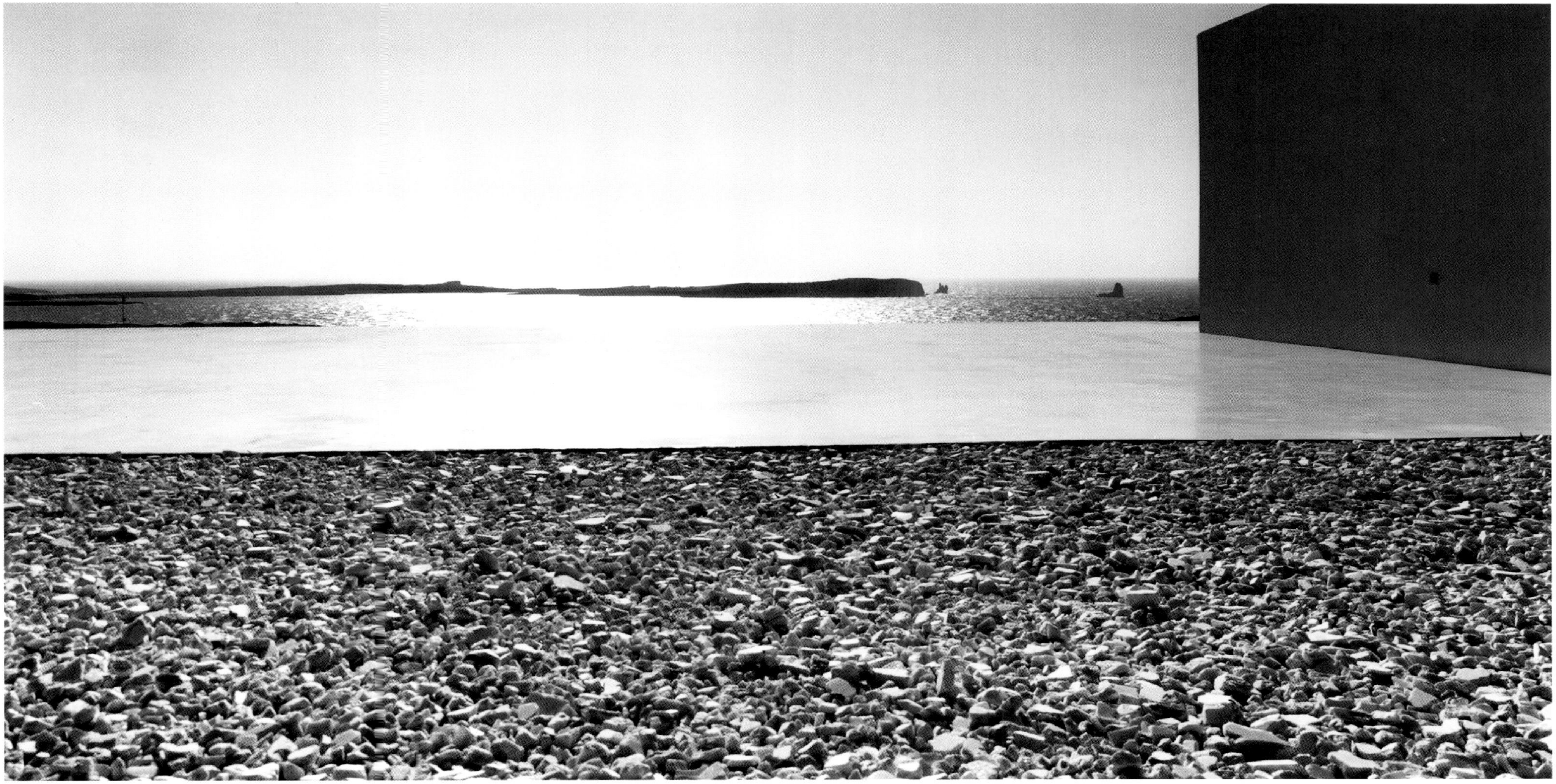

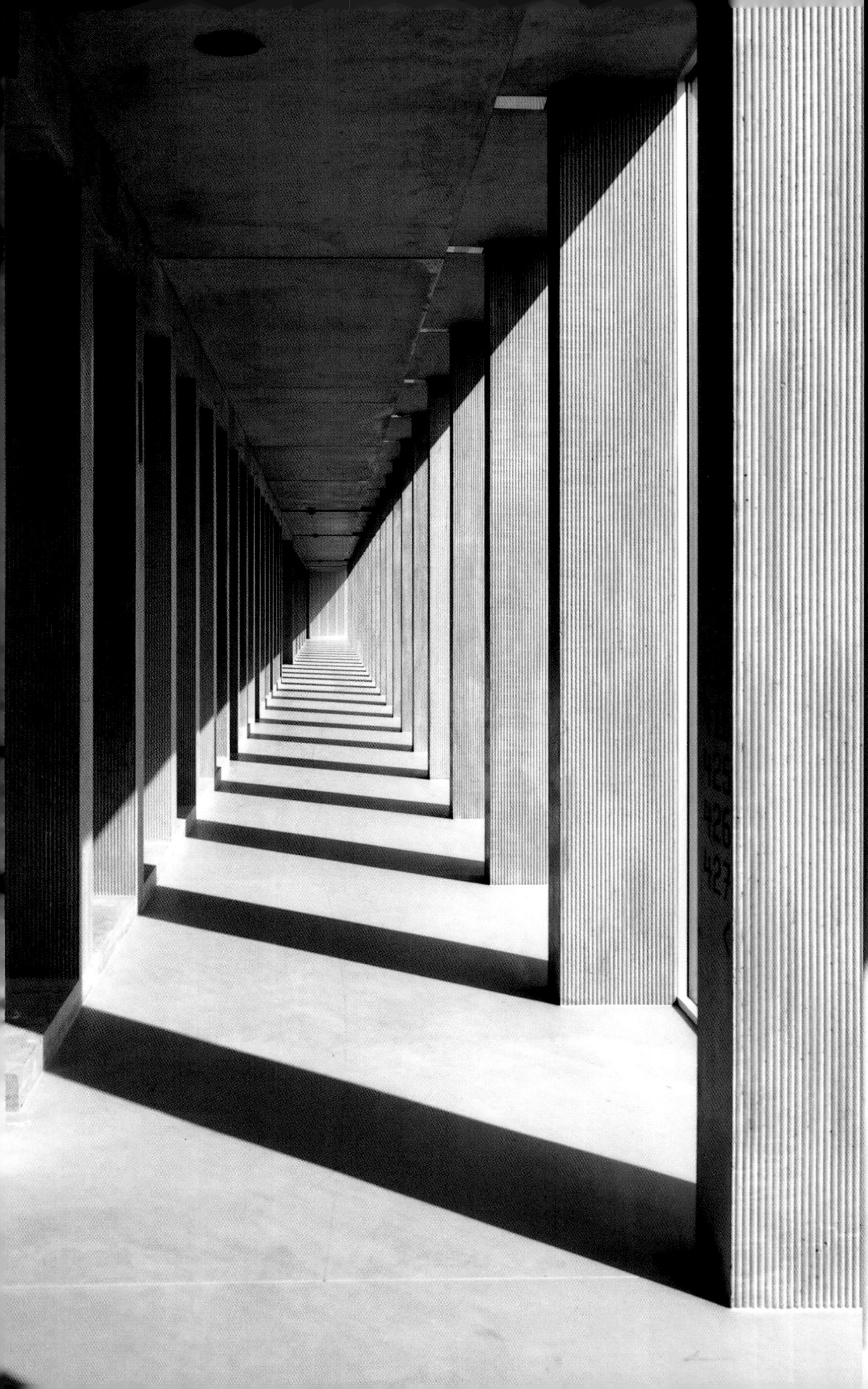

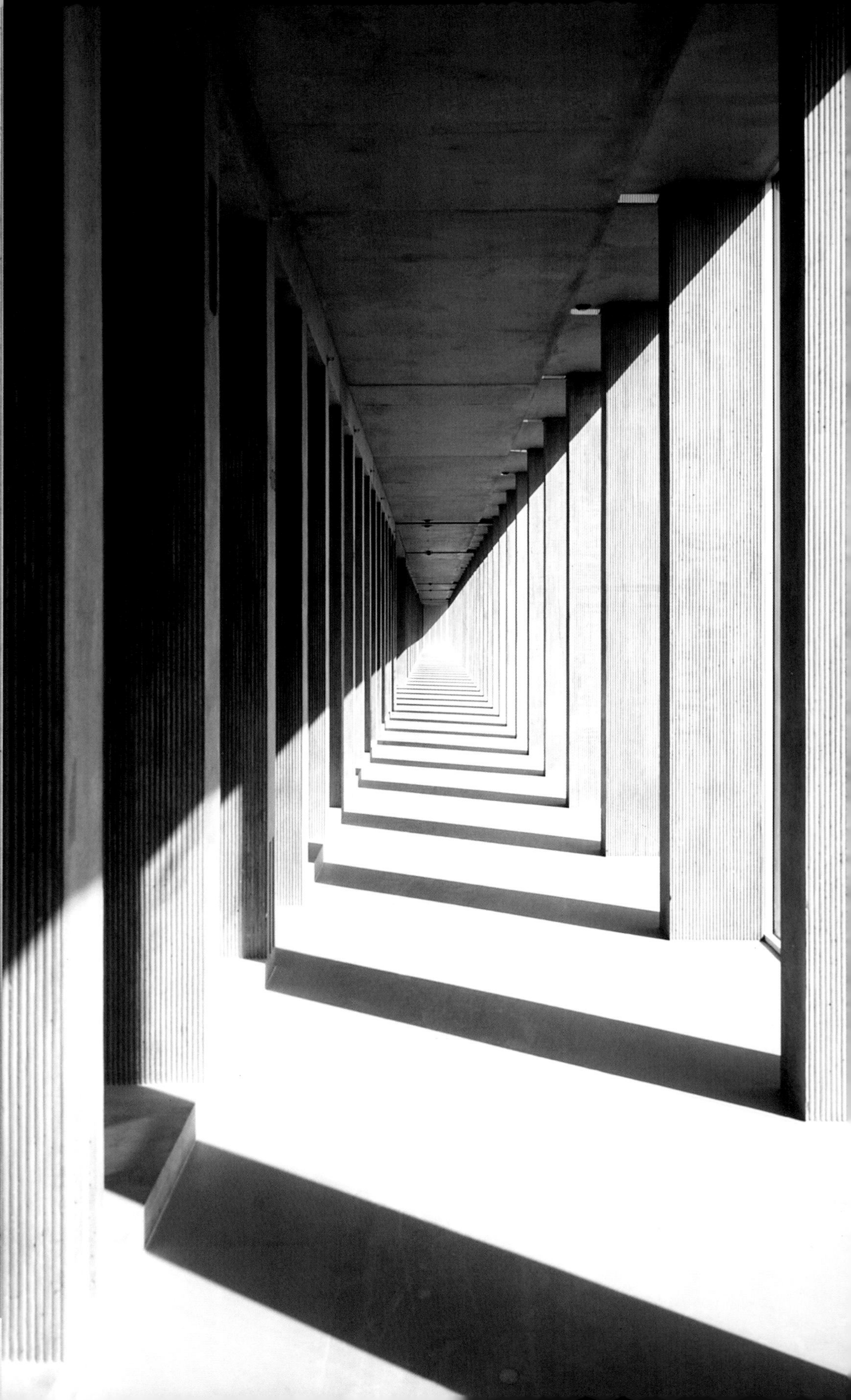

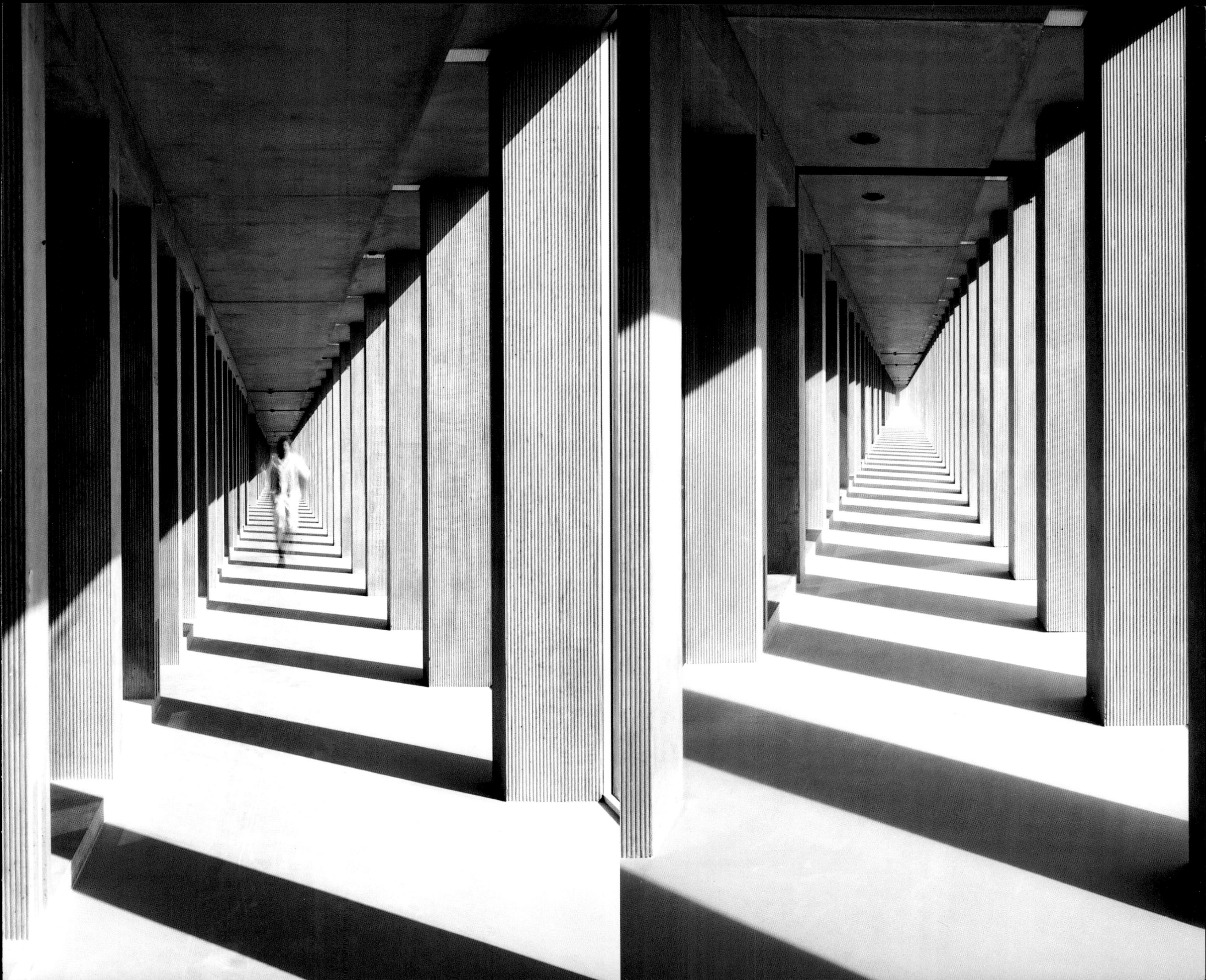

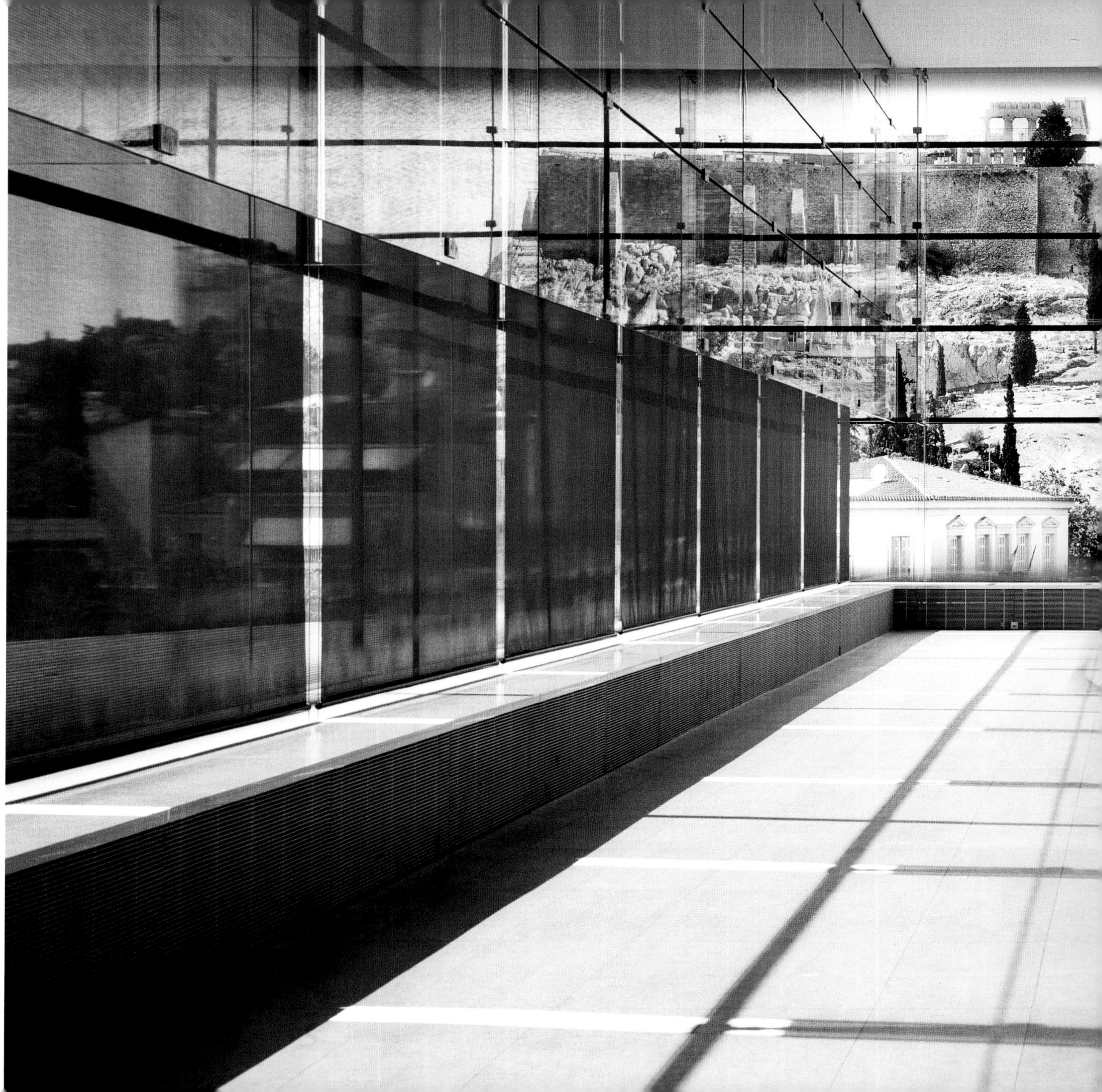

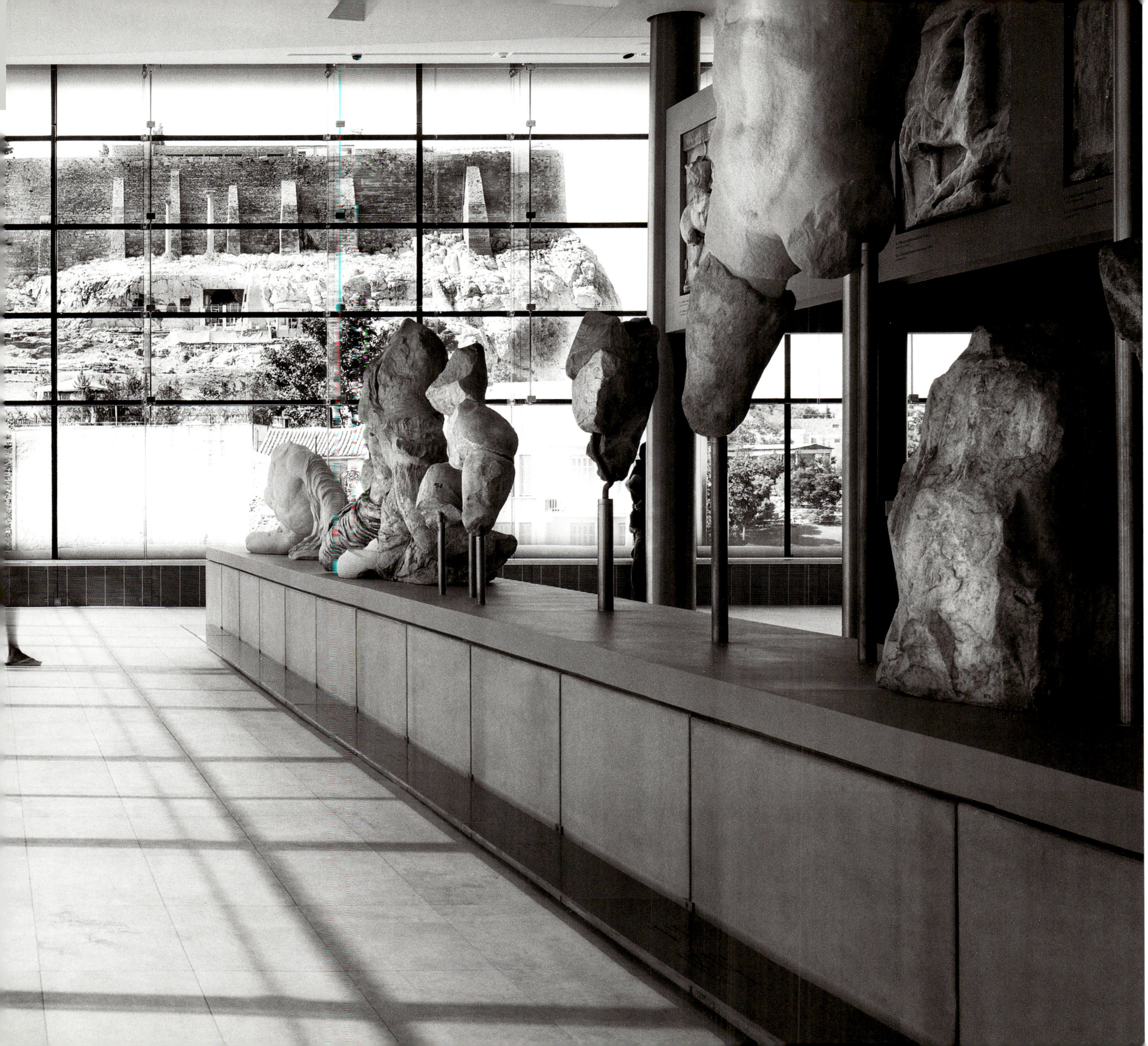

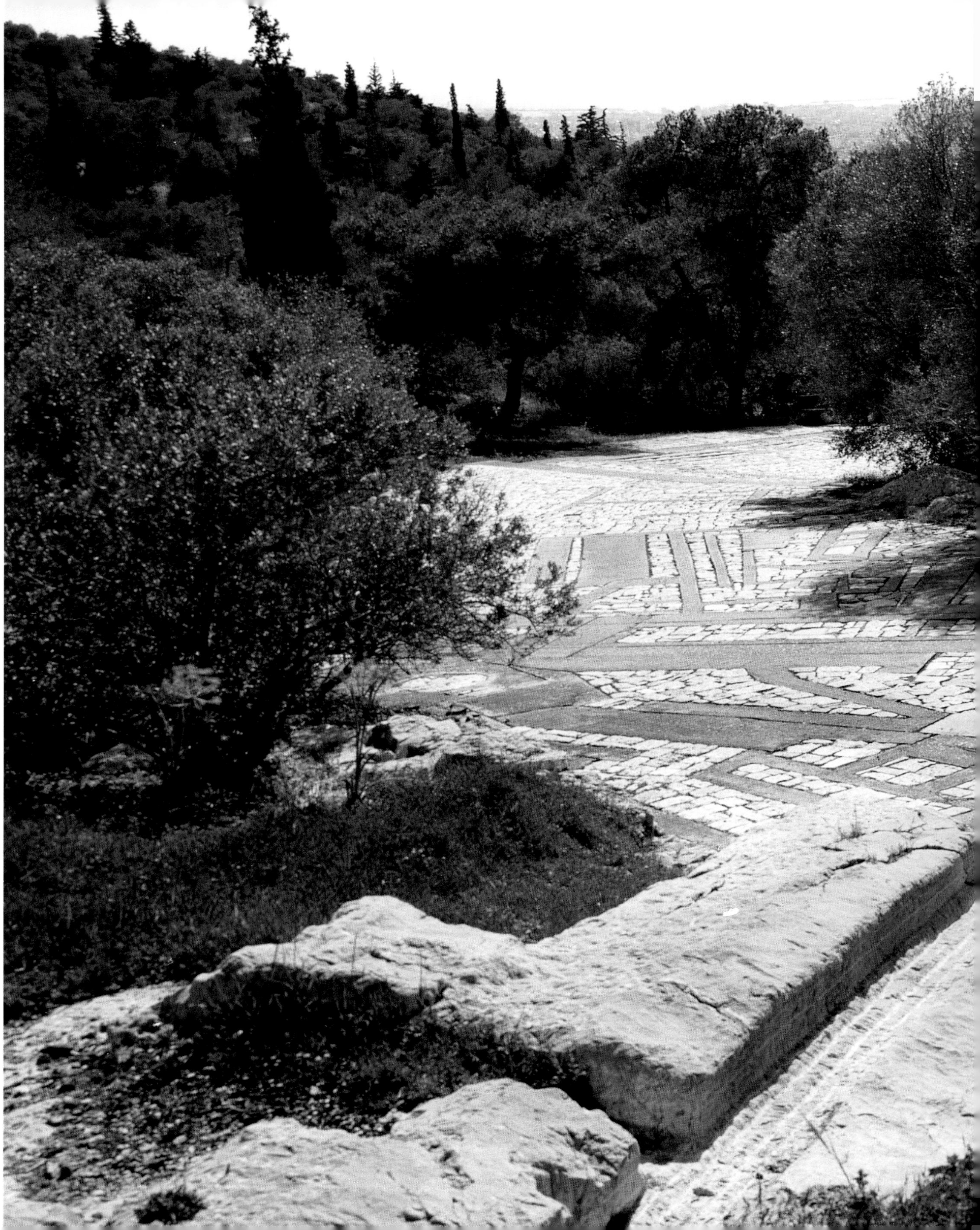

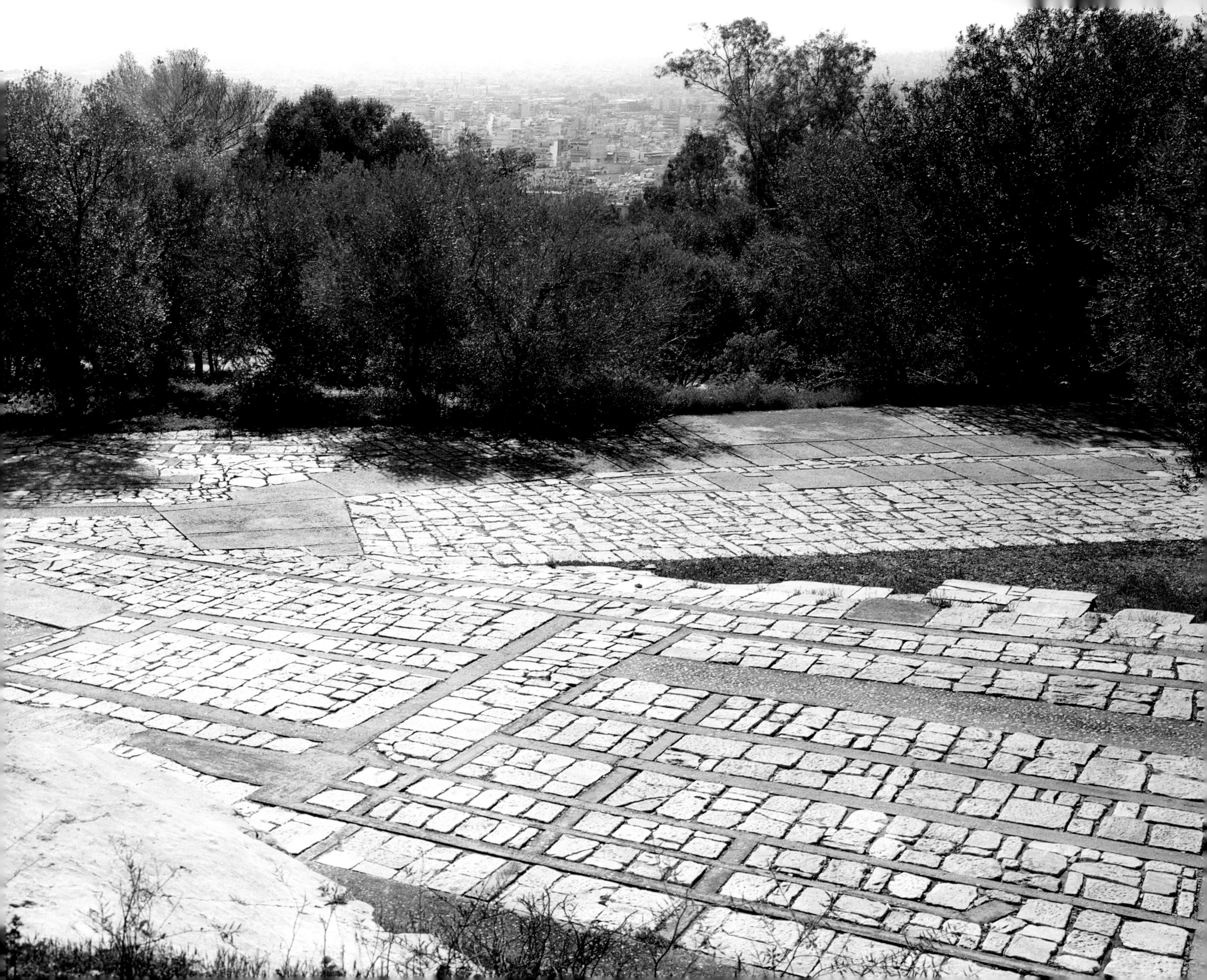

Crystal Silence

As a long distance runner during my teenage years I used to train daily in the pine forests of Chalki on Princes islands. It was during those long runs between the trees, the isolated Byzantine churches, and the Marmara Sea that I began to feel the urge to record the images I experienced on my solitary daily journey. It was then that I decided to become a photographer of monuments and landscapes.

Over the last decade I have walked endlessly through nature's timeless but ever changing landmarks. But I have also sought to see man's greatest architectural accomplishments, works which have sought to encompass natural beauty to their own vision.

And the voyage continues. The miles of the daily training of my teenage years have been multiplied into the millions and the world has finally become my new home island.

Erieta Attali, Santiago de Chile, 2010

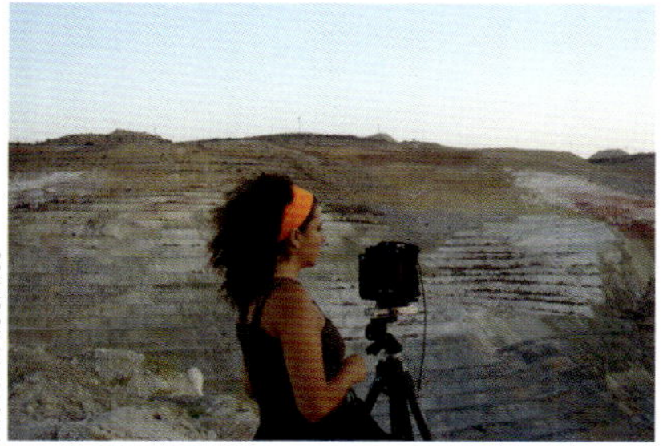

PHOTO: A. ROGAKOS

ERIETA ATTALI

Born in Tel Aviv, 1966, Erieta Attali studied photography at Goldsmith's College, University of London. From 1992 to 2002, she worked in the field of archaeological photography for institutions such as the British Museum, the National Gallery of London, the Archaeological Museum of Naples, and the Archaeological Museum of Izmir, among others.

Since beginning her career as an architectural photographer in 1998, her work has been published widely, both in monographs and in magazines such *Casabella, 2G, Quaderns, Harvard Design Magazine,* and *Bauwelt*. Her work has been the subject of numerous exhibitions, including the Benaki Museum, Athens, [2007] the Royal Danish Academy, Copenhagen [2005] the Royal Photographic Society, Bath, [1998] and the Museum of Finnish Architecture, Helsinki [1999]. In addition, Attali has been the recipient of numerous awards, including awards from the Fulbright Foundation, the Japan Foundation, and the Graham Foundation for Advanced Studies in the Fine Arts.

Since 2003, she has been an adjunct Assistant Professor of architectural photography at the Graduate School of Architecture, Planning and Preservation, Columbia University, New York. She is currently working on two projects: a photographic survey of Contemporary Scandinavian Architecture, in collaboration with Karin Skousbøll, and a monograph on contemporary architecture in South America.

ALESSIO ASSONITIS was born in Rome, graduated from Bennington College, and received his doctoral degree in art history from Columbia University. After serving as Allen Whitehall Clowes Curatorial Fellow at the Indianapolis Museum of Art, he arrived at the Medici Archive Project (MAP) in 2004 in Florence with a National Endowment for the Humanities Fellowship. He became MAP's Research Director in 2009, a position endowed by the Florence Gould Foundation. He has published scholarly work on Italian Renaissance painting, Mendicant aesthetics, antiquarian studies, Medici history, and archival studies.

KENNETH FRAMPTON was born in the United Kingdom in 1930 and trained as an architect at the Architectural Association School of Architecture, London. After practicing for a number of years in the United Kingdom and in Israel, he served as the editor of the British magazine Architectural Design. He has taught a number of leading institutions including the Royal College of Art, London, the ETH, Zurich, EPFL, Lausanne, the Accademia di Architettura, Mendrisio, and the Berlage Institute in The Netherlands. He is currently the Ware Professor of Architecture at the GSAPP, Columbia University, New York. He is the author of *Modern Architecture and the Critical Present* (1980), *Studies in Tectonic Culture* (1995), *American Masterworks* (1995), *Le Corbusier* (2001), *Labour, Work & Architecture* (2005), and an updated fourth edition of *Modern Architecture: A Critical History* (2007).

JUHANI PALLASMAA (b. 1936), architect SAFA, Hon. FAIA, Int FRIBA and professor, runs his architectural office in Helsinki. He has been active in urban planning, architecture, exhibition, product, and graphic design since early 1960s. He has held numerous academic positions, including Professor and Dean of the Helsinki University of Technology (1991–97); Director of the Museum of Finnish Architecture (1978–83), and Rector of the Institute of Design, Helsinki (1970–71). In addition to lecturing widely, he has held several visiting professorships in the United States, including the Raymond E. Maritz Visiting Professorship, Washington University in St. Louis (1999–2004); the Thomas Jefferson Visiting Professorship, University of Virginia (2002); the Eero Saarinen Visiting Professorship, Yale University (1993), and an Associate Professorship at the Haile Selassie I University, Addis Ababa (1972–74). Pallasmaa has published 26 books and over 300 essays in 30 languages, including *The Embodied Image*, (2010) *The Thinking Hand* (2009), *Encounters: Architectural Essays* (2005), *The Architecture of Image: Existential Space in Cinema* (2001 and 2007), and *The Eyes of the Skin: Architecture and the Senses* (1995 and 2005).

DIMITRI PHILIPPIDES graduated from the School of Architecture at the National Technical University (NTU) of Athens, Greece in 1962 and received his Arch.D. degree from The University of Michigan in 1973. He is professor emeritus of the School of Architecture, NTU of Athens, where he taught at the Department of Urban and Regional Planning (1975-2005). His professional work has extended from architectural design to urban planning. He has also written extensively on Greek architecture, traditional and modern, and on Greek physical planning. Aside from numerous articles, he has published fourteen books, edited two and translated another two.

JEANNETTE PLAUT received her architecture degree from the Universidad de Chile, as well as an architectural critic and academic. She works as a Latin American correspondent and has published her work in international journals such as *Architectural Record* (USA), *Metropolis* (USA), *Häuser* (Germany), *Taiwan Architecture Magazine*, *IW* (China), *Architektur Aktuell* (Austria), among others. She was the architecture editor of *Ambientes Magazine* (1998–2009), the Director of the National Exhibition of the XV Chilean Architecture Biennial (2006), Director of the International Architects Encounter of the XIV Chilean Architecture Biennial (2004) and Coordinator of the International Architects Encounter of the XIV Chilean Architecture Biennial (2002). She was the curator of the exhibition about architectonic and urban cases at the Chilean Pavilion at the International Exhibition Shanghai 2010. She worked as a consultant for the Sustainable Development and Human Settlements Division of ECLAC; United Nations (2004-2010) Among her recent publications there are: "PULSO: New Architecture in Chile" and a monograph on the Argentinian architect Rafael Iglesia. Currently she is working on a book series about contemporary Chilean architects. Plaut is the Director of PAL (Latin American Architecture Program) and editor in chief of *Trace Magazine* together with Marcelo Sarovic.

JILLY TRAGANOU (Ph.D., University of Westminster) is an architect, and Associate Professor in Spatial Design Studies at Parsons The New School for Design, New York. Her work examines social space and architecture in intersection with the fields of design studies and media studies, currently focusing on design's role in the configuration of national and postnational identities. She is the author of *The Tôkaidô Road: Traveling and Representation in Edo and Meiji Japan* (RoutledgeCurzon, 2004), and a co-editor of *Travel, Space, Architecture* (Ashgate, 2009) with Miodrag Mitrasinovic. She has published in the *Journal of Modern Greek Studies, Design Issues, Journal of Design History, L'architecture d'aujourd'hui, Journal for Architecture and Building Science of the Architectural Institute Japan,* and *Architecture in Greece,* and has lectured and participated in conferences internationally. She is currently working on the book *Designing the Olympics: Post-national Identity in the Age of Globalization* contracted by Routledge.

INDEX

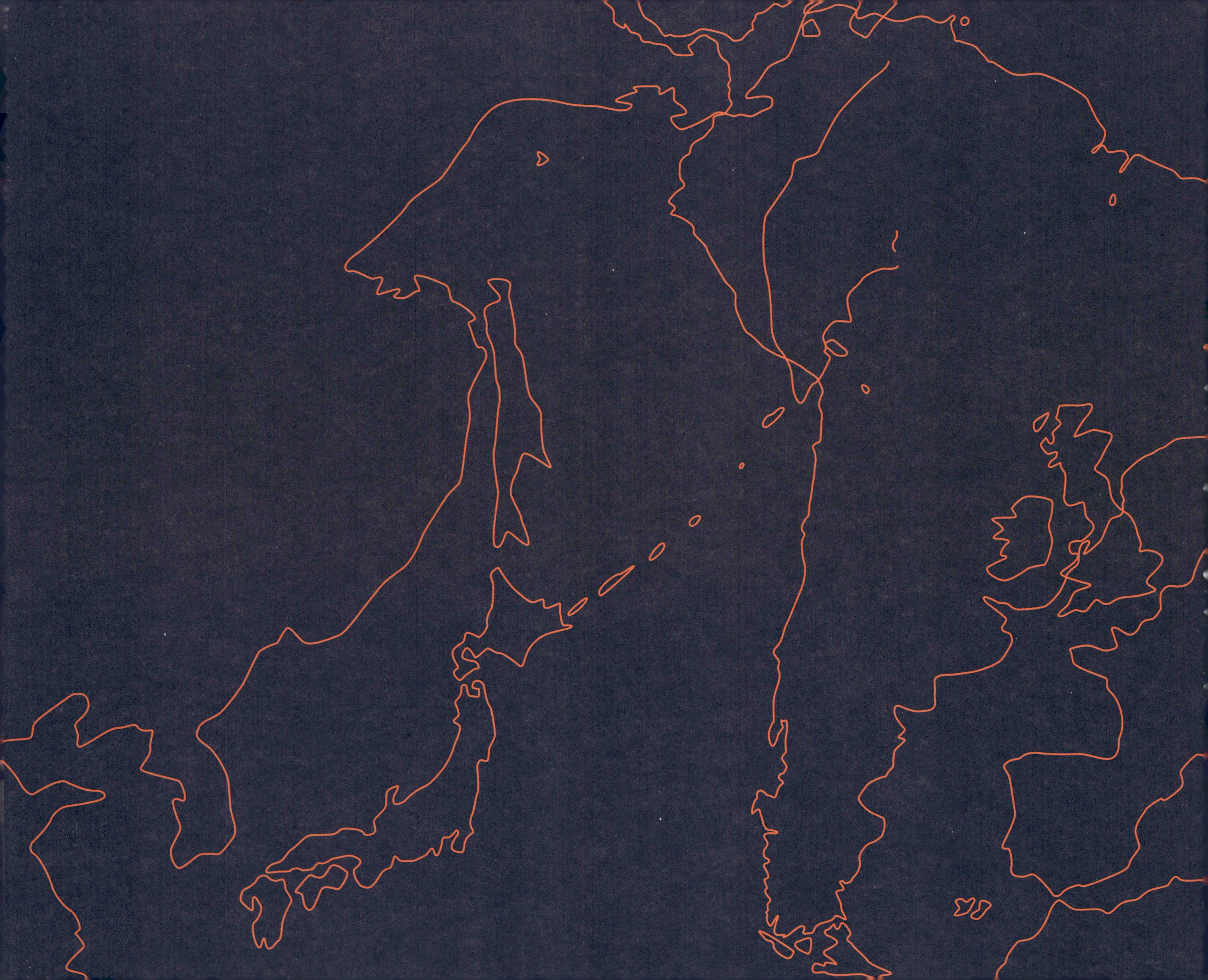